DIATOMIC MOLECULES
Results of *ab Initio* Calculations

DIATOMIC MOLECULES

Results of *ab Initio* Calculations

ROBERT S. MULLIKEN

and

WALTER C. ERMLER

THE UNIVERSITY OF CHICAGO
DEPARTMENT OF CHEMISTRY
CHICAGO, ILLINOIS

ACADEMIC PRESS New York San Francisco London 1977
A Subsidiary of Harcourt Brace Jovanovich, Publishers

ACADEMIC PRESS, INC.
111 Fifth Avenue, New York, New York 10003

United Kingdom Edition published by
ACADEMIC PRESS, INC. (LONDON) LTD.
24/28 Oval Road, London NW1

Library of Congress Cataloging in Publication Data

Mulliken, Robert Sanderson.
Diatomic molecules.

Includes bibliographical references and index.
1. Molecular theory. I. Ermler, Walter C., joint author. II. Title.
QD461.M78 541'.22 77-6605
ISBN 0-12-510750-1

PRINTED IN THE UNITED STATES OF AMERICA

CONTENTS

PREFACE

This small book began as a set of notes for use by chemical physics students in a course that aimed to illustrate the *results* obtained from wave-mechanical calculations on the electronic structure of first diatomic, and hopefully, then polyatomic molecules. We hope that the book may serve as a reference for researchers interested in the electronic structure of diatomic molecules as well as provide background analyses of related concepts for undergraduate and graduate students.

In Chapters II–VI, the main outlines of needed theory are presented as simply as possible. It is assumed that the reader has a background in the elements of quantum chemistry. Detailed theoretical derivations are not given except very briefly in Chapter I, which may be regarded as a theoretical introduction to the later chapters. One might at first glance at Chapter I, then go on to Chapter II.

In later chapters, the emphasis is on *ab initio* calculations by SCF (self-consistent-field) and multiconfiguration SCF molecular orbital methods. The approach is in terms of linear combination of atomic orbitals (LCAO) methods, with considerable emphasis on basis sets and on some details of configuration mixing to secure electron correlation. The plethora of other methods that have been and are being developed is mentioned only very briefly. Semiempirical calculations are not discussed.

In Chapter II, on one-electron molecules, several topics are introduced that are also relevant to later chapters but which can be well illustrated for the one-electron case: LCAO and LCMAO approximations (MAO, modified atomic orbitals) and basis sets, electronic population analysis, spectroscopic transition probabilities, and the nature of chemical bonding. In each succeeding chapter, new features of theory that become prominent when two or more electrons are present, or are important in hydrides, in homopolar molecules, or in heteropolar molecules, are successively introduced. Hence for a given topic in the Index, reference may be needed to more than one chapter.

The discussion and references are based largely on relatively recent papers, but basic earlier work is first considered in each chapter. The aim is to emphasize the best up-to-date work, through 1976. We apologize for references we may have overlooked. For a much more complete bibliography covering older work through 1973, see Richards *et al.* (Ref. 49 in Chapter III). No systematic attempt has been made to discuss all molecules on which *ab initio* calculations have been made. Rather, what has been presented is intended to be illustrative, although perhaps more comprehensive for heteropolar than for homopolar molecules.

We have recently become aware of a small book (R. F. W. Bader, "An Introduction to the Electronic Structure of Atoms and Molecules," Clarke, Irwin, & Co., Toronto, Vancouver, 1970) that complements ours in its clear explanation and presentation of contour maps of molecular charge distributions and of the differences between these and corresponding atomic distributions.

ACKNOWLEDGMENTS

We greatly appreciate permission from several authors to reproduce figures from their papers, and permission from the relevant journals. We are also very grateful to Professor Klaus Ruedenberg for letting us reproduce an unpublished figure (Fig. II–3). The book includes a number of tables not attributed to other authors but which were computed in this laboratory. We are grateful to Mr. Michael D. Allison for carrying out the calculations on population analysis. With reference to the rather numerous population analysis tables, we should emphasize that while they are usefully illustrative they are based on formulas that must in general be "taken with some grains of salt."

ACKNOWLEDGMENTS

LIST OF ACRONYMS

ANO	Approximate natural orbital
AO	Atomic orbital
APSG	Antisymmetrized product of strongly orthogonal geminals
CE	Correlation energy
CEPA	Coupled electron pair approximation
CI	Configuration interaction
CM	Configuration mixing
CSF	Configuration state function
GLF	Gaussian-lobe function
GTF	Gaussian-type function
GTO	Gaussian-type orbital
GVB	Generalized valence bond
HF	Hartree–Fock
IEPA	Independent electron pair approximation
INO	Iterative natural orbital
LCAO	Linear combination of atomic orbitals
LCGLF	Linear combination of Gaussian-lobe functions
LCGTF	Linear combination of Gaussian-type functions
LCMAO	Linear combination of modified atomic orbitals
LCSTF	Linear combination of Slater-type functions
MAO	Modified atomic orbital
MCSCF	Multiconfiguration self-consistent-field
MECE	Molecular extra correlation energy

MO	Molecular orbital
MSO	Molecular spin orbital
NO	Natural orbital
OVC	Optimized valence configurations
PNO	Pair (or pseudo) natural orbitals
POL-CI	Polarization configuration interaction
RHF	Restricted Hartree–Fock
SA	Separate atom
SAO	Separate-atom orbital
SASTF	Separate-atom Slater-type function
SCEP	Self-consistent electron pairs
SCF	Self-consistent field
SD	Slater determinant
STF	Slater-type function
STO	Slater-type orbital
UA	United atom
UAO	United-atom orbital
UASTF	United-atom Slater-type function
VCI	Valence configuration interaction

DIATOMIC MOLECULES
Results of *ab Initio* Calculations

CHAPTER I

INTRODUCTION

The main purpose of this book is to present a survey of the electronic structure of molecules as elucidated by means of *ab initio* quantum-mechanical calculations. New developments in the formal theory and the evolution of sophisticated computing facilities during the past two decades have led to innumerable important contributions to the basic understanding of molecular structure. The significance of these contributions will be exemplified in the discussion of results for representative molecules in the following chapters. We begin with a short review of the primary methods used in the computation of molecular wave functions and of related properties. The reader is referred to several representative sources for developments of the underlying quantum-mechanical theories [1]. Schaefer [2a, b] provides useful surveys of recent *ab initio* calculations, and has edited two volumes [2c] in a series on theoretical chemistry containing many excellent articles on methods of electronic structure theory. For an introductory discussion of diatomic spectra and structure, we suggest reference to Herzberg's well-known book [3].

The electronic structure of any molecule can be briefly characterized by giving an electronic configuration followed by a state symbol. The electron configuration consists of a listing of symbols for all the occupied molecular orbitals (MOs) in the order of decreasing strength of binding, with a superscript denoting the number of electrons in the given MO. For example, the ground-state electron configuration of N_2 is $1\sigma_g{}^2 1\sigma_u{}^2 2\sigma_g{}^2 2\sigma_u{}^2 1\pi_u{}^4 3\sigma_g{}^2$, while the state is ${}^1\Sigma_g{}^+$.

The MO symbols contain a serial number followed by a symbol for the *symmetry species*; each type of nuclear symmetry is represented by a different set of species symbols. For diatomic molecules, two types of symmetry exist—$D_{\infty h}$ for homopolar molecules and $C_{\infty v}$ for heteropolar molecules. The same symmetries occur for linear molecules in general.

For these molecules, the main species symbol indicates the value of the characteristic quantum number λ giving in units of $h/2\pi$ the magnitude $|m|$ of the orbital angular momentum $mh/2\pi$ around the symmetry axis; the symbols are σ, π, δ, φ, ... for $\lambda = 0, 1, 2, 3, \ldots$. For $D_{\infty h}$ molecules only, each symbol also contains a *parity* subscript which indicates the symmetry (g for even, u for odd) with respect to the operation of inversion of the wave function at the center of the molecule.

The state symbols are similar except that capital letters are used, which represent the *total* orbital angular momentum $\Lambda h/2\pi$ around the symmetry axis: Σ, Π, Δ, Φ, etc. However, there are two kinds of Σ states, Σ^+ and Σ^-, depending on whether the wave function does (Σ^-) or does not (Σ^+) change sign on reflection in a plane (any plane) passing through the nuclei. So we have $\Sigma_g{}^+$, $\Sigma_g{}^-$, $\Sigma_u{}^+$, $\Sigma_u{}^-$, Π_u, Π_g, Δ_g, Δ_u, and so on. The state symbols also are prefixed by a *multiplicity* superscript which indicates the quantum number of the resultant spin S (1, 2, 3, 4, ... for $S = 0, \frac{1}{2}, 1, \ldots$ respectively).

The detailed forms corresponding to the MO symbols are often given as LCAO (linear combination of atomic orbitals) expressions, or as linear combinations (LCSTF) of bits and pieces of AOs called STFs (Slater-type functions—see Section A). The AOs themselves have familiar symbols such as 1s, 2s, 2p, 3s, 3p, and 3d, but when used in building MOs, a symbol to indicate a particular value of λ must be added, as for example in 2pσ, 2pπ, 3dσ, 3dπ, and 3dδ. All s AOs are of type σ (or σ_g if the symmetry is $D_{\infty h}$) and one can write 1sσ, 2sσ, etc., but the σ can be understood without writing it explicitly. STFs can be symbolized in the same way as AOs, for example 1s, 2s, 2pσ, 2pπ, 2s, 3pσ, but it must then be understood that these symbols now refer in general to *pieces* of AOs, not to complete AOs. Instead of LCSTFs one can use LCGTFs or LCGLFs (see Section A). This can be done because each STF can be approximated by a linear combination of Gaussian-type functions (GTFs) or Gaussian-lobe functions (GLFs).

When one speaks of *the* electron configuration of a molecule, one is dealing with an approximation. For an exact wave function, other configurations must be mixed in.

The molecular wave functions from which the results here presented are obtained are of three general types:

(a) Hartree–Fock–Roothaan self-consistent-field (SCF) functions, each of which is an antisymmetrized product (Slater determinant) of one-electron functions called molecular spin orbitals (MSOs), each a product of a molecular orbital (MO) and a one-electron spin function; or in general a linear combination of such Slater determinants (SDs), called a configuration state function (CSF). Each CSF corresponds to a particular MO electron configuration, which means a specification of the number of electrons in each MO (here note that some MOs are degenerate); each MO is approximated as a linear combination of, usually, atomic basis functions;

(b) extensive linear combinations of CSFs corresponding to the superposition of different electronic configurations, commonly referred to as configuration interaction (CI) or, preferable, configuration mixing (CM)[1]; or

(c) more limited CM in the form of multiconfiguration self-consistent-field wave functions (MCSCF), which are obtained by the simultaneous optimization of the MOs of (a) and of CM coefficients corresponding to a selected (usually small) set of configurations. After an MCSCF function is obtained, it may be further improved by additional extensive CM.

A. BASIS SETS

The construction of good SCF MOs depends at present on having a well-chosen basis set.[2] Finite linear combinations of basis functions can then be used to approximate MOs to the desired level of accuracy within the limits of present-day computer technology. The two most commonly used kinds of basis functions are STFs and GTFs. The molecular basis set is usually formed by centering STFs or GTFs at each of the constituent atomic nuclei and/or sometimes at other points in regions near the nuclei. STFs are defined [6], in spherical polar coordinates, by

$$\chi_{nl}^{S} = N_{S} r^{n-1} \exp(-\zeta_{nl} r) Y_{lm}(\theta, \phi), \qquad (1)$$

where n, l, and m are principal, azimuthal, and magnetic quantum numbers, ζ_{nl} an "orbital exponent," $Y_{lm}(\theta, \phi)$ a spherical harmonic, and N a normalization factor such that

$$\langle \chi_{nl} | \chi_{nl} \rangle = \int_0^\infty \int_0^\pi \int_0^{2\pi} \chi_{nl}^* \chi_{nl} r^2 \sin\theta \, d\phi \, d\theta \, dr = 1. \qquad (2)$$

In Eq. (1) r is expressed in units of a_0 (atomic units). See Section II.C for further discussion.

[1] Hartree objected to the term "configuration interaction" and proposed "superposition of configurations" instead. We feel that Roothaan's term "configuration mixing" meets Hartree's objection but has the advantage of being much briefer.

[2] We should also mention recent investigations into the use of numerical methods for use in molecular calculations, see e.g. Refs. 4 and 5.

GTFs can be expressed [7] either in spherical polar coordinates as

$$\chi_{nl}^{\mathrm{Gs}} = N_{\mathrm{Gs}} r^{2n} \exp(-\zeta_{nl} r^2) Y_{lm}(\theta, \phi), \tag{3}$$

or in Cartesian coordinates as

$$\chi_{ijk}^{\mathrm{Gc}} = N_{\mathrm{Gc}} x^i y^j z^k \exp[-\zeta_{ijk}(x^2 + y^2 + z^2)]. \tag{4}$$

In Eqs. (1)–(4) the variables (r, θ, ϕ) or (x, y, z) measure the displacement of an electron from the point of reference of the basis function. GLFs are special cases of Eq. (3) where only the $\exp(-\zeta r^2)$ portion is used and several are centered at various positions in space in order to approximate the usual s, p, d, etc., atomic functions [8a]. When used in molecular wave functions GTFs are usually "contracted" [8], each of several GTFs being constrained to have fixed ratios of their coefficients, with the same ratios in each MO. The effective size of the basis set is thereby reduced, ideally with little loss of accuracy.

Experience has shown that extended basis sets used in the accurate computation of molecular wave functions are usually best set up by first adopting optimum basis sets previously obtained for the respective atoms [9], and adding "polarization functions," which are either functions having higher azimuthal quantum numbers than the occupied orbitals of the atoms [10] or are functions centered in regions of space other than at the atoms [11], or both. Calculations on Rydberg states of molecules in states with one or more MOs that are larger than in the ground state require additional basis functions with larger radial extents. For additional discussion on basis sets and on STFs, reference should be made to Section II.B.

B. HAMILTONIAN MATRIX ELEMENTS

A requirement common to both SCF–MO and CM procedures is the evaluation of integrals or matrix elements of the basis functions with respect to the terms in the nonrelativistic Hamiltonian. (Up to now, nearly all calculations have been nonrelativistic.) For a molecule having A nuclei of charge Z_a and N electrons, the Hamiltonian, in atomic units (electronic mass m_e, electronic charge e, and $h/2\pi$ set to unity), is

$$\mathscr{H} = -\frac{1}{2}\sum_{\mu}^{N} \nabla_{\mu}^{2} - \sum_{\mu}^{N}\sum_{k}^{A} \frac{Z_k}{r_{\mu k}} + \sum_{\mu<\nu}^{N} \frac{1}{r_{\mu\nu}} + \sum_{a<b}^{A} \frac{Z_a Z_b}{r_{ab}}, \tag{5}$$

where we have assumed that the nuclei are fixed in space. The terms in Eq. (5) correspond, respectively, to the kinetic ($\hat{T}$), nuclear attraction ($\hat{V}^{\mathrm{na}}$), electronic repulsion ($\hat{G}$), and nuclear repulsion ($\hat{V}^{\mathrm{nr}}$), operators. The molecular wave function satisfying the Schrödinger equation

$$\mathscr{H}\Psi = E\Psi, \tag{6}$$

being an eigenfunction of $\mathscr{H}$, yields the total energy of the molecule as its eigenvalue. Assuming that Ψ is normalized, we have

$$E = \langle\Psi|\hat{\mathscr{H}}|\Psi\rangle = \langle\Psi|\hat{T}|\Psi\rangle + \langle\Psi|\hat{V}^{\text{na}}|\Psi\rangle + \langle\Psi|\hat{G}|\Psi\rangle + V^{\text{nr}}. \quad (7)$$

(The nuclear repulsion energy is a function only of internuclear distances, which are assumed to be fixed.)

The initial step in the SCF–MO and CM methods is the computation of integrals with respect to the basis functions and the $\hat{T}$, $\hat{V}^{\text{na}}$, and $\hat{G}$ operators. It is well known that, because of certain permutational symmetries, a basis set comprised of m functions leads to $(m^2+m)/2$ $\hat{T}$ and $\hat{V}$ ("one-electron") matrix elements and to $(m^4+2m^3+3m^2+2m)/8$ $\hat{G}$ ("two-electron") matrix elements. These quantities are defined as

$$T_{pq} = \langle\chi_p|\hat{T}|\chi_q\rangle, \quad (8)$$

$$V_{pq}^{\text{na}} = \langle\chi_p|\hat{V}^{\text{na}}|\chi_q\rangle, \quad (9)$$

$$G_{pqrs} = \langle\chi_p\chi_q|\hat{G}|\chi_r\chi_s\rangle, \quad (10)$$

where the subscripts p, q, r, and s label the basis functions χ, which may be centered at one or more nuclei or at various points in space. A third set of one-electron integrals needed for the analysis comprise the overlap matrix

$$S_{pq} = \langle\chi_p|\chi_q\rangle. \quad (11)$$

The matrix elements defined in Eqs. (8)–(11) are machine-computed[3] and stored for use in the subsequent SCF–MO and CM procedures.

C. THE HARTREE–FOCK–ROOTHAAN EQUATIONS

MO theory was catapulted into extensive quantitative use with the development of the matrix Hartree–Fock self-consistent field equations by Roothaan in 1951 [12]. If it is assumed that the state of each electron in a molecule may be represented by an MSO, the total wave function for an N-electron molecular state which contains only closed shells of MOs is constructed as an antisymmetrized product, or Slater determinant, of MSOs,

$$\Psi = \mathscr{A}(\psi_1 \quad \psi_2 \quad \cdots \quad \psi_N) = (N!)^{-1/2}\begin{vmatrix} \psi_1(1) & \psi_2(1) & \cdots & \psi_N(1) \\ \psi_1(2) & \psi_2(2) & \cdots & \psi_N(2) \\ \vdots & \vdots & & \vdots \\ \psi_1(N) & \psi_2(N) & \cdots & \psi_N(N) \end{vmatrix} \quad (12)$$

[3] Several efficient integral evaluation programs are available from the Quantum Chemistry Program Exchange, Indiana University.

If we now assume that each MO ϕ may be approximated as a linear combination of m basis functions χ_p (STFs or GTFs) and that there are $n = N/2$ doubly occupied MOs, then

$$\phi_i = \sum_p^m \chi_p c_{pi} \tag{13}$$

or in matrix notation

$$\boldsymbol{\phi} = \boldsymbol{\chi}\mathbf{c}, \tag{14}$$

where $\boldsymbol{\phi}$ is a column vector of dimension n, $\boldsymbol{\chi}$ a row vector of dimension m, and $\mathbf{c}$ a matrix of m rows and n columns. It is also required that the MOs be orthonormal,

$$\langle \phi_i | \phi_j \rangle = \delta_{ij}, \tag{15}$$

or

$$\boldsymbol{\phi}^\dagger \boldsymbol{\phi} = \mathbf{c}^\dagger \mathbf{S} \mathbf{c} = \mathbf{1}, \tag{16}$$

where $\mathbf{S}$ is the overlap matrix [Eq. (11)], δ_{ij} the Kronecker delta, $\mathbf{1}$ the unit matrix, and † denotes the adjoint operation.

The total energy with respect to the MOs ϕ_i is now [Eq. (7)]

$$E^{\mathrm{SCF}} = 2\sum_i^n h_i + \sum_{i,j}^n (2J_{ij} - K_{ij}), \tag{17}$$

where

$$h_i = \langle \phi_i | \hat{T} + \hat{V}^{\mathrm{na}} | \phi_i \rangle, \tag{18}$$

$$J_{ij} = \langle \phi_i(1)\,\phi_j(2) | \frac{1}{r_{12}} | \phi_i(1)\,\phi_j(2) \rangle, \tag{19}$$

$$K_{ij} = \langle \phi_i(1)\,\phi_j(2) | \frac{1}{r_{12}} | \phi_j(1)\,\phi_i(2) \rangle. \tag{20}$$

In Eqs. (19) and (20) the J_{ij} are designated as Coulomb and K_{ij} as exchange repulsion integrals relative to electrons 1 and 2.

To form the SCF equations the J_{ij} and K_{ij} are first rewritten in terms of the one-electron Coulomb ($\hat{J}_i$) and exchange ($\hat{K}_i$) operators:

$$\hat{J}_i(2)\,\phi_j(2) = \langle \phi_i(1) | \frac{1}{r_{12}} | \phi_i(1) \rangle\, \phi_j(2), \tag{21}$$

$$\hat{K}_i(2)\,\phi_j(2) = \langle \phi_i(1) | \frac{1}{r_{12}} | \phi_j(1) \rangle\, \phi_i(2). \tag{22}$$

The first-order variation of the total energy combined with the orthonormality constraint imposed through the Lagrangian multipliers ε_{ij} leads to the Hartree–Fock–Roothaan SCF equations [12]

$$(\mathbf{F} - \boldsymbol{\varepsilon}\mathbf{S})\,\mathbf{c} = 0, \tag{23}$$

where the Fock operator $\hat{F}$ is defined by

$$\hat{F} = \hat{h} + \sum_i (2\hat{J}_i - \hat{K}_i). \tag{24}$$

The total molecular wave function is obtained by solving Eq. (23); that is, by finding the solution of the secular equation

$$\det(\mathsf{F} - \boldsymbol{\varepsilon}\mathsf{S}) = 0. \tag{25}$$

The roots or eigenvalues $\boldsymbol{\varepsilon}$ of Eq. (23) are called the orbital energies and the matrix c that diagonalizes F consists of the corresponding eigenvectors c_i. Hence, the problem of determining the wave functions for a closed-shell molecule comprised of N electrons is reduced to the straightforward solution of a secular equation involving the basis set expansion coefficients c [Eqs. (13) and (14)]. In practice, an initial guess of c is chosen and Eq. (23) is solved; a new Fock operator [Eq. (24)] is formed using the first improved c and Eq. (23) is solved again. This procedure is repeated until the total energy and eigenvectors are unchanged to the extent of some predetermined threshold.

The SCF energy, when cast into the form

$$E^{\mathrm{SCF}} = \sum_i^n e_i + \sum_i^n \varepsilon_i, \tag{26}$$

where

$$e_i = \langle \phi_i | \hat{h} | \phi_i \rangle, \tag{27}$$

$$\varepsilon_i = e_i + \sum_j (2J_{ij} - K_{ij}), \tag{28}$$

leads to an interesting diagrammatic representation. Take the Mg atom as an example, with

$$\Psi^{\mathrm{SCF}}(\mathrm{Mg}) = \mathscr{A}[1s\alpha(1)\,1s\beta(2)\,2s\alpha(3)\,2s\beta(4)\,2p_{+1}\alpha(5)\,2p_{+1}\beta(6)\; 2p_0\alpha(7)\,2p_0\beta(8)\,2p_{-1}\alpha(9)\,2p_{-1}\beta(10)\,3s\alpha(11)\,3s\beta(12)]; \tag{29}$$

the SCF energy is then given by summing the elements of Table 1. Note that the sum over the indices i and j includes every $i \neq j$ twice, so that only half the J_{ij} and K_{ij} entries are needed in E^{SCF}. The various hole states of Mg^+ can be obtained by the removal of one ε_i column, as is illustrated in the table; this shows that $\varepsilon_i \approx I_i$ (Koopmans' theorem [13]). This result is not exact because

(a) the SCF orbitals of Mg^+ are somewhat smaller than those of the corresponding Mg MOs, and

(b) E^{SCF} is not the exact E of either Mg or Mg^+;

both have corrections for correlation energy.

Several authors [14] have reformulated the closed-shell Hartree–Fock–Roothaan equations to apply to certain classes of open-shell configurations;

TABLE 1

Nonrelativistic Mg Atom Energy Terms in SCF Approximation Omitting Magnetic Terms[a]

	$1s\alpha$	$1s\beta$	$2s\alpha$	$2s\beta$	$2p_{+1}\alpha$	$2p_{+1}\beta$	$2p_0\alpha$	$2p_0\beta$	$2_{p-1}\alpha$	$2_{p-1}\beta$	$3s\alpha$	$3s\beta$
$1s\alpha$	e_k	J_{kk}	$(J-K)_{kl}$	J_{kl}	$(J-K)_{kp}$	J_{kp}	$(J-K)_{kp}$	J_{kp}	$(J-K)_{kp}$	J_{kp}	$(J-K)_{km}$	J_{km}
$1s\beta$		e_k	J_{kl}	$(J-K)_{kl}$	J_{kp}	$(J-K)_{kp}$	J_{kp}	$(J-K)_{kp}$	J_{kp}	$(J-K)_{kp}$	J_{km}	$(J-K)_{km}$
$2s\alpha$			e_l	J_{ll}	$(J-K)_{lp}$	J_{lp}	$(J-K)_{lp}$	J_{lp}	$(J-K)_{lp}$	J_{lp}	$(J-K)_{lm}$	J_{lm}
$2s\beta$				e_l	J_{lp}	$(J-K)_{lp}$	J_{lp}	$(J-K)_{lp}$	J_{lp}	$(J-K)_{lp}$	J_{lm}	$(J-K)_{lm}$
$2p_{+1}\alpha$					e_p	J_{pp}	$J_{pp}-K_{+0}$	J_{pp}	$J_{pp}-K_{+-}$	J_{pp}	$(J-K)_{pm}$	J_{pm}
$2p_{+1}\beta$						e_p	J_{pp}	$J_{pp}-K_{+-}$	J_{pp}	$J_{pp}-K_{+-}$	J_{pm}	$(J-K)_{pm}$
$2p_0\alpha$							e_p	J_{pp}	$J_{pp}-K_{-0}$	J_{pp}	$(J-K)_{pm}$	J_{pm}
$2p_0\beta$								e_p	J_{pp}	$J_{pp}-K_{-0}$	J_{pm}	$(J-K)_{pm}$
$2p_{-1}\alpha$								J_{pp}	e_p	J_{pp}	$(J-K)_{pm}$	J_{pm}
$2p_{-1}\beta$								$J_{pp}-K_{-0}$		e_p	J_{pm}	$(J-K)_{pm}$
$3s\alpha$								J_{pm}			e_m	J_{km}
$3s\beta$								$(J-K)_{pm}$				e_m

[a] AOs are all orthonormal (can use $2p_x$, $2p_y$, $2p_z$ ($\equiv 2p\sigma$) instead of $2p_{+1}$, $2p_{-1}$, and $2p_0$).

that is, those electronic states which have some MOs singly occupied. As a simple example we write the expression for the total energy of a configuration containing either a single open shell or certain cases of several open shells [14a]. If the wave function is divided into a closed-shell (doubly occupied) part and an open-shell (singly occupied) part and we assign labels k, l to orbitals in the former set and m, n to those in the latter set, then

$$E = 2\sum_{k} h_k + \sum_{k,l}(2J_{kl} - K_{kl})$$
$$+ f\left(2\sum_{m} h_m + f\sum_{m,n}(2aJ_{mn} - bK_{mn}) + 2\sum_{k,m}(2J_{km} - K_{km})\right). \quad (30)$$

In Eq. (30) f is a fractional occupation number corresponding to the ratio of the number of occupied to the number of available open-shell orbitals and a and b are coefficients [14a, b, g] specific to the state under consideration. The appropriate SCF equations are derived [14a] from Eq. (30) by defining various *coupling operators* and carrying out a variational procedure similar to that used for the closed-shell case [12].

More elaborate analyses [14b–g] are required for systems whose total energy cannot be represented by Eq. (30). In practice, the total wave function of an electronic state is represented by a CSF; that is, the single Slater determinant or the fixed linear combination of Slater determinants that have the correct spatial and spin symmetries. The expectation value of the nonrelativistic Hamiltonian [Eq. (7)] is then formed to define the various coupling coefficients, and, finally, the SCF equations are derived and solved by the basis set expansion method.

Although the restricted or conventional Hartree–Fock (RHF) method previously discussed has been used most extensively, several other approaches which allow for the removal of various symmetry and equivalence restrictions [15] imposed on the space and spin portions of the RHF orbitals while retaining their independent particle forms have been developed. Since only RHF, CM, and MCSCF types of calculations will be discussed, we will not present a detailed summary of these "extended" HF procedures. Several of these methods are reviewed in Ref. 2a.

D. CONFIGURATION MIXING (CM)

The wave functions derived from the Hartree–Fock method are composed of antisymmetrized products of single-particle spin orbitals (Slater determinants) in which the many-particle effects are approximated by the interaction of each electron with an average field produced by all the remaining electrons in the system. The error introduced by this self-consistent field approximation is known as the correlation energy. Operationally it has been defined [16] as the difference between the exact solution of the nonrelativistic Hamiltonian and its expectation value in the Hartree–Fock approximation. Although the correlation energy is usually only a small percentage of the total molecular energy, it is usually of the same order of magnitude as the strengths of chemical bonds and of electronic excitation energies. Hartree–Fock wave functions determined for molecules near equilibrium also may not produce the correct separated or united-atom electronic states, and as a result will not yield good potential energy surfaces for bond dissociation.

The most common method of improving the Hartree–Fock approximation is through the use of configuration mixing (CM). The total molecular wave function is represented as a sum of many-particle functions which may be called configuration state functions (CSFs) [16–18]. Each CSF is either a single SD (for closed-shell and certain special open-shell systems) or a specific linear combination of SDs possessing the proper symmetry of the electronic state under consideration; that is, the CSFs are eigenfunctions of all spin and space operators that commute with the Hamiltonian [18a]. The CM wave function may be written

$$\Psi = \sum C_I \Phi_I, \tag{31}$$

where I indexes the CSFs Φ_I and the coefficients C_I are determined by applying the variation principle to the energy expectation value. This leads to the secular equation

$$(\mathsf{H} - E\mathbf{1})\mathbf{C} = 0, \tag{32}$$

where the matrix H is made up of the elements

$$H_{IJ} = \langle \Phi_I | H | \Phi_J \rangle; \tag{33}$$

E and $\mathbf{C}$ are the respective eigenvalue and eigenvector of H for the desired electronic state and $\mathbf{1}$ is the unit matrix.

Operationally, CM calculations are usually done in several stages. First, all appropriate one- and two-electron basis function integrals are computed [Eqs. (8)–(11)], and MOs are obtained by some type of SCF, MCSCF, or approximate natural orbital (ANO) [19, 20] calculation. The integrals, which are usually expressed relative to STFs, GTFs, or GLFs (or symmetry adapted linear combinations of these), are then transformed in terms of the MOs chosen to represent the configuration space. These are usually the occupied plus some or all the virtual MOs arising from the SCF or MCSCF calculation or the ANOs. This transformation produces the MO basis functions needed for the construction of the CSFs already discussed. Next, the Hamiltonian matrix elements [Eq. (33)] with respect to the CSFs are computed. Finally, the desired eigenstates of the system are determined from Eq. (32).[4] A possible additional step involves the construction of the one-

[4] Techniques for doing CM calculations without constructing H have been developed [21].

electron density matrix that may be used in the generation of natural orbitals [19, 20] and the computation of one-electron properties (see below).

The CSFs are usually characterized by the number of electrons excited from the MOs of a chosen reference configuration; that is, CSFs corresponding to the excitation of one electron from the reference CSF are called singly excited, those having two electrons excited are called doubly excited, etc. For a system of N electrons represented by a configuration space comprised of m basis functions, the number of CSFs is approximately m^N if all excitations are considered [22]. Another rule of thumb is that the number of configurations is nearly proportional to m^i, where $i = 1, 2, 3$, etc., for singly, doubly, triply, etc., excited configurations. This indicates that CM calculations involving large basis sets, many electrons, and many configurations will be prohibitive unless some approximations are made; for example, the numbers of CSFs due to all single and double excitations of the ten valence electrons relative to the Hartree–Fock configurations of ${}^1\Sigma_g{}^+$ and ${}^3\Pi_g$ states of N_2 relative to a basis of five s-, four p-, and three d-type STFs ($m = 44$) are 2670 and 22,333 respectively. Some methods currently being used for selecting configurations will be discussed.

E. MULTICONFIGURATION SELF-CONSISTENT FIELD (MCSCF) WAVE FUNCTIONS

As in the CM approach, the total wave function for the system of interest is a linear combination of CSFs. In MCSCF methods, however, both the linear coefficients c_{pi} of the basis functions x_{pi} that comprise the orbitals ϕ_i [Eq. (13)] *and* the mixing coefficients C_I for the CSFs that comprise the total wave function Ψ [Eq. (31)] are simultaneously optimized variationally with respect to the total energy. Since there are several excellent articles [23] on the formalism utilized in MCSCF methods, we present here only a simplified outline of the method.

The MCSCF equations are derived by writing the expression for the total electronic energy for the multiconfiguration wave function and then applying the variation principle.

To illustrate the formalism briefly, we follow the treatment of Veillard and Clementi [23a]. The total electronic wave function is written as

$$\Psi = a_{00}\Phi_{00} + \sum_{t=1}^{n} \sum_{n=1}^{w-n} a_{tu}\Phi_{tu}. \tag{34}$$

Equation (34) refers to a closed-shell system having $2n$ electrons distributed among n doubly occupied orbitals $\phi_1, \ldots, \phi_n$ [the "(n)" set] and a second set of orbitals $\phi_{n+1}, \ldots, \phi_w$ [the "$(w-n)$" set]. The ϕ_i are defined in terms of the basis functions χ_p and expansion coefficients c_{pi} as in Eq. (13). Consider all possible double excitations from the (n) to the $(w-n)$ set, where a given excitation from the former to the latter is designated $t \rightarrow u$ with t ranging from 1 to n and u from $n+1$ to w. Thus, in Eq. (35) and in equations to follow, the t or t' indexes the (n) set and u or u' the $(w-n)$ set.

The total electronic energy for Ψ of Eq. (35) is [23a]

$$E = 2\sum_{t=1}^{n}\left[h_t + \sum_{t'=1}^{n} P_{tt'} - A_t\left(h_t + 2\sum_{t'=1}^{n} P_{tt'} - P_{tt}\right)\right]$$
$$+ \sum_{u=1}^{w-n} 2B_u\left(h_u + P_{uu} + 2\sum_{t=1}^{n} P_{tu}\right) + 2\sum_{t=1}^{n}\sum_{n=1}^{w-n} a_{tu}(a_{00}K_{tu} - 2a_{tu}P_{tu})$$
$$+ \sum_{t,t'=1}^{n} A_{tt'}K_{tt'}(1-\delta_{tt'}) + \sum_{u,u'=1}^{w-n} B_{uu'}K_{uu'}(1-\delta_{uu'}), \tag{35}$$

where

$$a_{00}^2 + \sum_{t=1}^{n}\sum_{u=1}^{w-n} a_{tu}^2 = 1, \tag{36}$$

$$A_{tt'} = \sum_{u=1}^{w-n} a_{tn}a_{t'n} \qquad (A_t \equiv A_{tt}), \tag{37}$$

$$B_{uu'} = \sum_{t=1}^{n} a_{tu}a_{tu'} \qquad (B_u \equiv B_{uu}), \tag{38}$$

and h_i and $P_{ij} \equiv J_{ij} - \frac{1}{2}K_{ij}$ refer to the usual one-electron, Coulomb, and exchange matrix elements [see Eqs. (18)–(20)].

The relationships

$$1 = a_{00}^2 + \sum_{t=1}^{n} A_t = a_{00}^2 + \sum_{u=1}^{w-n} B_u \tag{39}$$

or

$$\sum_{t=1}^{n} A_t = \sum_{u=1}^{w-n} B_u \tag{40}$$

imply that the coefficient A_t represents the fraction of an electron that is excited from the ϕ_t orbital of the (n) set to the totality of the ϕ_u orbitals of the $(w-n)$ set. Conversely, the coefficient B_u represents the fraction of an electron in the ϕ_u orbital due to excitation from the entire (n) set. If Φ_{00} has the "zeroth-order configuration" ${\phi_1}^2{\phi_2}^2 \cdots {\phi_n}^2$, the complete MCSCF wave function [Eq. (34)] will be a sum of $(wn-n^2)$ zeroth-order configurations. Thus, A_t and B_t coefficients may be used to summarize the overall configuration as

$$\phi_1^{2(1-A_1)} \cdots \phi_n^{2(1-A_n)}\phi_{n+1}^{2B_1} \cdots \phi_w^{2B_{w-n}}.$$

To obtain the best orbitals ϕ and mixing coefficients a_{tu}, the variation principle is applied to Eq. (35) by requiring

$$\left(\frac{\partial E}{\partial \phi_t}\right) = \left(\frac{\partial E}{\partial \phi_u}\right) = \left(\frac{\partial E}{\partial a_{00}}\right) = \left(\frac{\partial E}{\partial a_{tu}}\right) = 0.$$

The necessary equations are given in terms of the operators

$$F_t = (1-A_t)h_t + 2\sum_{t'=1}^{n}(1-A_t-A_{t'})P_t + 2A_tP_t + 2\sum_{u=1}^{w-n}B_uP_u$$

$$+\sum_{u=1}^{w-n}(a_{00}a_{tu}K_u - 2a_{tu}^2P_u) + \sum_{t'=1}^{n}A_{tt'}K_{t'}(1-\delta_{tt'}), \tag{41a}$$

$$F_u = B_u\left(h_u + 2P_u\right) + \sum_{t=1}^{n}(a_ta_{00}K_t - 2a_{tu}^2P_t) + \sum_{u'=1}^{w-n}B_{uu'}K_{u'}(1-\delta_{uu'}). \tag{41b}$$

After carrying out the above differentiations and imposing the orthogonality constraints

$$\langle\phi_i|\phi_j\rangle = \delta_{ij}, \tag{42}$$

where i and j run over the full (n) and $(w-n)$ sets, the orbitals ϕ_t and ϕ_u are given by

$$F_t\phi_t - \sum_{t'\neq t=1}^{n}\phi_{t'}\theta_{tt'} - \sum_{u=1}^{w-n}\phi_u\theta_{ut} = \phi_t\theta_{tt}, \tag{43a}$$

$$F_u\phi_u - \sum_{u'\neq u=1}^{w-n}\phi_{u'}\theta_{uu} - \sum_{t=1}^{n}\phi_t\theta_{tu} = \phi_u\theta_{uu}, \tag{43b}$$

where the θ's are Lagrangian multipliers. Using the definitions

$$T_t = \sum_{t'\neq t=1}^{n}|\phi_{t'}\rangle\langle\phi_{t'}|F_t|\phi_t\rangle, \tag{44a}$$

$$T_u = \sum_{u=1}^{w-n}|\phi_u\rangle\langle\phi_u|F_t|\phi_t\rangle, \tag{44b}$$

$$U_u = \sum_{u'\neq u=1}^{w-n}|\phi_{u'}\rangle\langle\phi_{u'}|F_u|\phi_u\rangle, \tag{44c}$$

$$U_t = \sum_{t=1}^{n}|\phi_t\rangle\langle\phi_t|F_u|\phi_u\rangle, \tag{44d}$$

we can rewrite Eqs. (43) as

$$|F_t - T_t - T_u|\phi_t\rangle = |\phi_t\rangle\theta_{tt}, \tag{45a}$$

$$|F_u - U_u - U_t|\phi_u\rangle = |\phi_u\rangle\theta_{uu}. \tag{45b}$$

The equations defining the mixing coefficients a_{tu} are similarly obtained by coupling the conditions

$$0 = a_{00}\,\delta a_{00} + \sum_{t=1}^{n}\sum_{u=1}^{w-n} a_{tu}\,\delta a_{tu}, \tag{46}$$

$$\delta A_t = 2\sum_{u=1}^{w-n} a_{tu}\,\delta a_{tu}, \tag{47a}$$

$$\delta B_t = 2\sum_{t=1}^{n} a_{tu}\,\delta a_{tu} \tag{47b}$$

into the first-order variation of Eq. (35) through the Lagrangian multipliers $\lambda/2$. The results are secular equations for the mixing coefficients

$$(E_{00}-E)\,a_{00} + \sum_{t=1}^{n}\sum_{u=1}^{w-n} a_{tu}\,K_{tu} = 0, \tag{48}$$

$$(E_{tu}-E)\,a_{tu} + a_{00}\,K_{tu} + \sum_{t'}\sum_{u'} a_{t'u'}(K_{tt'}\,\delta_{uu'}\,K_{uu'}\,\delta_{tt'})(1-\delta_{uu'}\,\delta_{tt'}) = 0, \tag{49}$$

where $\lambda = E_0 - E$ has been taken into account. Equations (45), (48), and (49) are MCSCF equations that may be used to determine the optimum orbitals ϕ and mixing coefficients a_{tu} for Ψ of Eq. (34). The explicit dependence of the basis function expansion coefficients c_{pi} is seen [23a] by a straightforward substitution of Eq. (13) for the orbitals ϕ. In practice, an MCSCF calculation proceeds by making an initial guess of the expansion and mixing coefficients c_{pi} and a_{tu} and Eqs. (45) are solved for new ϕ's; the secular equations (48) and (49) are then solved for new a_{tu}'s. These are used to construct the matrix operators necessary for a subsequent solution of Eqs. (45). The resultant orbitals are substituted into Eqs. (48) and (49) to compute improved a_{tu}'s, and the procedure is iterated until convergence is achieved in both the orbitals and mixing coefficients to predetermined thresholds.

F. DENSITY MATRICES AND NATURAL ORBITALS

At this juncture it is useful to point out the importance of density matrices and natural orbitals in molecular structure calculations. The reader is referred to Löwdin [24], McWeeny [1b], Ruedenberg [25], and Davidson [26] for more extensive discussions of these topics. We present here a brief summary and some examples of the properties of density matrices and natural orbitals.

The first-order density matrix $\rho(1,1')$ for an N-electron electronic eigenfunction is defined as

$$\rho(1,1') = N\int \Psi(\tau_1,\tau_2,\tau_3,\ldots,\tau_N)\,\Psi^*(\tau_1',\tau_2,\tau_3,\ldots,\tau_N)\,d\tau_2\,d\tau_3\cdots d\tau_N, \tag{50}$$

where τ_i may include space coordinates x_i and spins η_i. Each Ψ is in general a linear combination of Slater determinants. If desired, the spins can be eliminated from Eq. (50) by first setting $\eta_i = \eta_i'$ and integrating over all $d\eta_i$. Assuming this done, it can be shown that $\rho(1, 1')$ in general takes the form

$$\rho(1, 1') = \sum_{i,j} \gamma_{ij} \phi_j(1) \phi_i^*(1'), \tag{51}$$

where 1 and 1′ refer to the space coordinates of electron 1, and the ϕ's are MOs (or AOs in the case of atoms). For correlated wave functions, the ϕ's include both SCF occupied orbitals and corresponding virtual orbitals (not STFs or GTFs, although of course the orbitals may be built from these; but that is irrelevant for the present discussion). Equation (51) can be represented in matrix form as

$$\rho(1, 1') = \operatorname{tr} \left\{ \begin{pmatrix} \gamma_{11} & \gamma_{12} & \cdots & \gamma_{1N} \\ \gamma_{21} & \gamma_{22} & \cdots & \gamma_{2N} \\ \vdots & & & \\ \gamma_{N1} & \cdot & \cdots & \gamma_{NN} \end{pmatrix} \begin{pmatrix} \phi_1(1) \\ \phi_2(1) \\ \vdots \\ \phi_N(1) \end{pmatrix} (\phi_1^*(1') \phi_2^*(1') \cdots \phi_N^*(1')) \right\}$$

$$= \operatorname{tr} \Gamma\Phi\Phi'. \tag{52}$$

In simple situations, only terms with $i = j$ are present, but in general, nondiagonal terms also appear. Valuable results are then obtained by first diagonalizing the γ matrix in the usual way, thereby finding its eigenvalues λ_k and the corresponding eigenvectors U_k—the former form a diagonal matrix Λ and the latter a unitary matrix U. We now set up a unitary transformation of the matrix product $\Gamma\Phi\Phi'$ in Eq. (52) to which its trace is invariant. Thus,

$$\rho(1, 1') = \operatorname{tr} U^\dagger \Gamma\Phi\Phi' U = \operatorname{tr} U^\dagger \Gamma U U^\dagger \Phi\Phi' U = \operatorname{tr} \Lambda(U^\dagger\Phi)(\Phi' U). \tag{53}$$

The matrix product $U^\dagger\Phi$, or alternatively $\Phi' U$, yields a new set of basis orbitals, say χ_i (not to be confused with the basis functions χ of Sections (A–C), which are the "*natural orbitals*" (NOs) of the problem,

$$U^\dagger\Phi = \begin{pmatrix} U_{11} & U_{21} & \cdots & U_{N1} \\ U_{12} & U_{22} & \cdots & U_{N2} \\ \vdots & & & \\ U_{1N} & \cdot & \cdots & U_{NN} \end{pmatrix} \begin{pmatrix} \phi_1(1) \\ \phi_2(1) \\ \vdots \\ \phi_N(1) \end{pmatrix} = \begin{pmatrix} \chi_1(1) \\ \chi_2(1) \\ \vdots \\ \chi_N(1) \end{pmatrix},$$

$$\Phi' U = \chi_1^*(1') \chi_2^*(1') \cdots \chi_N^*(1'). \tag{54}$$

Thus from Eq. (53),

$$\rho(1, 1') = \lambda_1 \chi_1(1) \chi_1^*(1') + \lambda_2 \chi_2(1) \chi_2^*(1') + \cdots + \lambda_N \chi_N(1) \chi_N^*(1'). \tag{55}$$

Now set $1' = 1$; $\rho(1,1)$ then gives the charge density in terms of the *occupation numbers* λ_k of the N natural orbitals χ_k.

Examples (all MOs assumed orthonormal)

(1) For H_2, let

$$\Psi = \frac{1}{\sqrt{2!}}\begin{vmatrix} 1\sigma_g\alpha(1) & 1\sigma_g\beta(1) \\ 1\sigma_g\alpha(2) & 1\sigma_g\beta(2) \end{vmatrix};$$

$$\rho(1,1') = 1\sigma_g(1)\,1\sigma_g(1')$$

after eliminating spin.

(2) For LiH, let

$$\Psi = \frac{1}{\sqrt{4!}}\begin{vmatrix} 1\sigma\alpha(1) & 1\sigma\beta(1) & 2\sigma\alpha(1) & 2\sigma\beta(1) \\ 1\sigma\alpha(2) & 1\sigma\beta(2) & 2\sigma\alpha(2) & 2\sigma\beta(2) \\ 1\sigma\alpha(3) & 1\sigma\beta(3) & 2\sigma\alpha(3) & 2\sigma\beta(3) \\ 1\sigma\alpha(4) & 1\sigma\beta(4) & 2\sigma\alpha(4) & 2\sigma\beta(4) \end{vmatrix};$$

$$\begin{aligned}\rho(1,1') &= 2[1\sigma(1)\,1\sigma(1') + 2\sigma(1)\,2\sigma(1')] \\ &= \operatorname{tr}\left\{\begin{pmatrix} 2 & 0 \\ 0 & 2 \end{pmatrix}\begin{pmatrix} 1\sigma(1) \\ 2\sigma(1) \end{pmatrix}(1\sigma(1')\,2\sigma(1'))\right\}.\end{aligned}$$

With a single determinant as here and in Example (1), Γ is already diagonal and nothing new is obtained. The SCF MOs are automatically NOs.

(3) For LiH^+, let

$$\Psi = a1\sigma^2(1,2)\,2\sigma\alpha(3) + b1\sigma^2(1,2)\,3\sigma\alpha(3);$$

$$\begin{aligned}\rho(1,1') &= 2(a^2+b^2)\,1\sigma(1)\,1\sigma(1') + a^2\,1\sigma(1)\,2\sigma(1') + b^2\,3\sigma(1)\,3\sigma(1') \\ &\quad + ab2\sigma(1)\,3\sigma(1') + ab3\sigma(1)\,2\sigma(1') \\ &= \operatorname{tr}\left\{\begin{pmatrix} 2(a^2+b^2) & 0 & 0 \\ 0 & a^2 & ab \\ 0 & ab & b^2 \end{pmatrix}\begin{pmatrix} 1\sigma(1) \\ 2\sigma(1) \\ 3\sigma(1) \end{pmatrix}(1\sigma(1')\,2\sigma(1')\,(3\sigma'))\right\},\end{aligned}$$

which after diagonalizing Γ can be reduced to

$$\rho(1,1') = \lambda_1\chi_1(1)\chi_1(1') + \lambda_2\chi_2(1)\chi_2(1') + \lambda_3\chi_3(1)\chi_3(1')$$

with $\lambda_1 = 2$, $\chi_1 \approx 1\sigma$, $\lambda_2 + \lambda_3 = 1$. The generalization to $\Psi = \sum_1^n C_i\,1\sigma^2 n\sigma\alpha$ is obvious.

(4) $\Psi = a1\sigma^2 2\sigma + b2\sigma^2 3\sigma$;

$$\rho(1,1') = 2a^2\,1\sigma(1\sigma') + (a^2+2b^2)\,2\sigma(1)\,2\sigma(1') + b^2\,3\sigma(1)\,3\sigma(1').$$

There are no nondiagonal terms in Γ, and $\rho(1,1')$ cannot be reduced; for reduction to be possible, different determinants can differ by only one MSO—as in Example (3).

(5) $\Psi = a1\sigma_g^2 + b[1\sigma_g(1)2\sigma_g(2)+2\sigma_g(1)1\sigma_g(2)] + c2\sigma_g^2;$

$$\rho(1,1') = (a^2+b^2)1\sigma_g(1')1\sigma_g(1) + (ab+bc)[1\sigma_g(1')2\sigma_g(1) + 2\sigma_g(1')1\sigma_g(1)] + c^2 2\sigma_g(1')2\sigma_g(1)$$

$$= \mathrm{tr}\left\{\begin{pmatrix} a^2+b^2 & ab+bc \\ ab+bc & c^2 \end{pmatrix}\begin{pmatrix} 1\sigma_g(1) \\ 2\sigma_g(1) \end{pmatrix} 1\sigma_g(1')2\sigma_g(1')\right\}$$

yielding $\lambda_1 1\sigma_g'(1)1\sigma_g'(1')+\lambda_2 2\sigma_g'(1)2\sigma_g'(1')$ on reduction, where $1\sigma_g'$ and $2\sigma_g'$ are NOs which are linear combinations of the original $1\sigma_g$ and $2\sigma_g$. This implies that although the coefficients in the original Ψ must have been variationally determined, the functions $1\sigma_g$ and $2\sigma_g$, somehow specified, were not NOs. Note further that an MCSCF treatment of Ψ in the form originally given would also have yielded the form $A(1\sigma_g)^2+B(2\sigma_g)^2$ expressed in terms of NOs. The preceding discussion is readily generalized to show that in terms of NOs,

$$\Psi = A(1\sigma_g)^2 + B(1\sigma_u)^2 + C(2\sigma_g^2) + D(1\pi_u^2) + \cdots.$$

This result cannot, however, be generalized for systems having more than two electrons.

The usefulness of natural orbitals for computations on many-electron systems has been, primarily, to reduce the number of configurations to be employed by means of a truncation of the orbital basis function space as represented by some approximate natural orbitals.

It should be emphasized that the natural orbitals χ of Eq. (55) are strictly obtainable only *after* the density matrix $\rho(1,1')$ has been determined. Since the natural orbitals are related to the orbitals ϕ used to construct $\rho(1,1')$ by a unitary transformation, they will necessarily yield the same energy and properties, although they will be of a form more amenable to physical interpretation. This means that since, in practice, the exact natural orbitals χ cannot be derived until the exact "canonical" orbitals $\boldsymbol{\phi}$ have already been determined they are of no use in reducing the computational effort in a many-electron calculation. A recent study by Shavitt *et al.* [27] compares the effect of using several orbital transformations on the convergence of configuration mixing expansions (Section D).

One of the earliest applications of natural orbital transformations to computations on diatomic molecules was developed by Bender and Davidson [28]. Called the iterative natural orbital (INO) method, it involves five steps:

(a) making an initial guess of the natural orbitals in terms of a basis set,

(b) constructing an approximate wave function from some selected set of configurations formed from these orbitals,

(c) computating the natural orbitals for this wave function,

(d) using these natural orbitals as a guess, and

(e) repeating steps (b) and (c) until the wave function and orbitals are stable to within some threshold.

Although the INO procedure is nonvariational, it has proven to be reasonably convergent in many cases.

Another promising method for exploiting the convergence properties of natural orbitals has been suggested by several authors [29]. It involves the computation of orbitals using a truncated set of configurations as determined using perturbation theory [22] or some other *a priori* selection scheme, followed by a transformation of the resultant wave function to natural orbitals, then a truncation of these orbitals omitting those with the smallest occupation numbers λ. The truncated orbitals are then used to define a new, more complete, set of configurations, which is significantly smaller than would have resulted had the entire original basis set been employed. The Hamiltonian matrix due to these new orbital and configuration spaces is then solved by standard methods (Section D). Although this procedure could be iterated as already described, it is found that the initial reduction of the orbital space allows for the inclusion of most of the required added configurations.

G. ELECTRON PAIR METHODS

A major bottleneck in the use of configuration mixing techniques is the inherently slow convergence of the expansion of a molecular wave function in terms of CSFs [18]. To illustrate this difficulty the wave function is expressed as a sum of terms, each of which corresponds to a certain level of excitation,

$$\Psi = C_0 \Phi_0 + \sum_{I,A} C_I{}^A \Phi_I{}^A + \sum_{I<J,\, A<B} C_{IJ}^{AB} \Phi_{IJ}^{AB} + \cdots, \tag{56}$$

where the CSFs $\Phi_I{}^A, \Phi_{IJ}^{AB}, \ldots$ represent single, double, etc., substitutions of electrons from the occupied $(I, J, \ldots)$ to the unoccupied or virtual $(A, B, \ldots)$ orbitals, and Φ_0 is either the SCF solution or a small set of dominant reference CSFs obtained from, for example, an MCSCF calculation. As was mentioned in Section D, the number of CSFs in each summation in Eq. (56) is nearly proportional to M^i, where M is the number of orbitals and i is the level of excitation. It is clear from Eq. (56) that the inclusion of double, triple, and higher excitations soon results in excessive numbers of CSFs.

The single replacements, represented by the $C_I{}^A$ of Eq. (56) can be made to vanish by Brillouin's theorem if the MSO basis is properly chosen (all $\Phi_I^A \equiv 0$). They can also be eliminated by an appropriate choice of MSOs such that the coefficients C_I^A are zero. The MSOs are usually termed "Brillouin" or "Brueckner" orbitals, respectively, depending on which of the above choices is made. The total energy may then be expressed [30],

$$E = E_0 + \sum_{I<J} \varepsilon_{IJ}, \tag{57}$$

where E_0 is the energy due to $C_0 \Phi_0$ and the ε_{IJ} are the so-called pair correlation energies.

The possibility of expressing the total energy as a sum of pair energies has been suggested by several authors [31, 32]. The earliest attempts to compute the ε_{IJ} were based on the use of orbitals termed antisymmetrized products of strongly orthogonal geminals (APSG) [33, 34] Φ_k^R for $R = 1, 2, \ldots, n/2$ pairs of electrons and $k = 1, 2, \ldots, m$ orbitals such that

$$\langle \Phi_k^R | \Phi_l^S \rangle = \delta_{RS}\, \delta_{kl}. \tag{58}$$

The APSGs are said to account only for intrapair correlation effects. The important consequence of strong orthogonality is that the n-electron system is represented by a sum of localized pairs and the variational procedure carried out on coefficients of the doubly substituted APSGs. Quadruple and higher substitutions may be approximated as products of doubles, termed "unlinked clusters" [35]. Methods that are represented by the form Eq. (57) are usually termed independent electron pair approximations (IEPA) [30]. The IEPA may also include interpair terms, that is, singlet and triplet coupled excitations from different geminals. It should also be noted that IEPA energies are in general nonvariational.

Having a wave function that is expressed as a sum of pair interaction terms lends itself to the application of natural orbital transformations as outlined in Section F. Edmiston and Krauss [36] showed that in the framework of the IEPA, natural orbitals could be computed for the separate pairs and subsequently employed in a wave function of the form Eq. (56) in a conventional configuration mixing procedure. Since these are not the true natural orbitals which would be derived from a total density matrix for a wave function of the system taken as a whole, they were termed pseudonatural orbitals (PNOs). The designation "pair natural orbitals" is used interchangeably with pseudonatural orbitals and is generally recognized as having the same meaning.

The PNOs derived from an IEPA procedure can be used with standard configuration mixing (Section D) to get new mixing coefficients for the wave function (normalized such that the coefficient of Φ_0 is unity),

$$\Psi_{\text{CI}} = \Phi_0 + \sum_{A<B,\, I<J} C_{IJ}^{AB}\, \Phi_{IJ}^{AB}, \tag{59}$$

that will ensure a variational result for the energy. Meyer [37] has formulated a PNO–CI procedure whereby spin orbitals ϕ_a and ϕ_b, into which substitutions are made, are NOs of the IEPA geminals. An important feature is the relaxation of the orthogonality constraint, viz. whereas the Φ_{IJ}^{AB} are mutually orthogonal, the PNOs of different geminals are nonorthogonal since for each Φ_{IJ}^{AB} the density matrix for the pair is put into diagonal form and the orbitals suitably rotated into NOs independently of the remaining pairs.

The coefficients in the expansion for Ψ_{CI} of Eq. (59) are obtained as eigenvectors of the Hamiltonian matrix for m pairs,

$$\mathsf{H}^{\text{CI}} = \begin{pmatrix} E_0 & \mathsf{H}_{01} & \cdots & \mathsf{H}_{0m} \\ \mathsf{H}_{10} & \mathsf{H}_{11} & \cdots & \mathsf{H}_{1m} \\ \vdots & \vdots & & \vdots \\ \mathsf{H}_{m0} & \mathsf{H}_{m1} & \cdots & \mathsf{H}_{mm} \end{pmatrix} = \mathsf{H}_{\mu\nu}, \tag{60}$$

where E_0 is a constant and the block index μ refers to the number of pairs [38]. The eigenvectors of H are of the same blocked form

$$\mathsf{C}^{\text{CI}} = (1, \mathbf{C}_1^{\text{CI}}, \ldots, C_m^{\text{CI}}). \tag{61}$$

The m coupled eigenvalue equations for the PNO–CI method are

$$\begin{pmatrix} E_0 & \mathsf{H}_{0\mu} \\ \mathsf{H}_{\mu 0} & \mathsf{H}_{\mu\mu} \end{pmatrix} \begin{pmatrix} 1 \\ \mathsf{C}_\mu^{\text{CI}} \end{pmatrix} - E_{\text{CI}} \begin{pmatrix} 1 \\ \mathsf{C}_\mu^{\text{CI}} \end{pmatrix} = - \sum_{\nu \neq \mu = 1}^{m} \begin{pmatrix} \mathsf{H}_{0\nu} \mathsf{C}_\nu^{\text{CI}} \\ \mathsf{H}_{\mu\nu} \mathsf{C}_\mu^{\text{CI}} \end{pmatrix}. \tag{62}$$

These may be compared with the IEPA equations

$$\begin{pmatrix} E_0 & H_{0\mu} \\ \mathsf{H}_{\mu 0} & H_{\mu\mu} \end{pmatrix} \begin{pmatrix} 1 \\ \mathsf{C}_\mu^{\text{IEPA}} \end{pmatrix} - (E_0 + \varepsilon_\mu^{\text{IEPA}}) \begin{pmatrix} 1 \\ \mathsf{C}_\mu^{\text{IEPA}} \end{pmatrix} = 0. \tag{63}$$

Equations (63) may be used in an iterative procedure to obtain the C^{CI} of Eq. (62) with the C^{IEPA} as first approximations [38].

To facilitate the inclusion of unlinked clusters, Meyer [37] has proposed the use of what he terms the coupled-electron-pair approximation (CEPA). This extension of the PNO–CI method may be described by writing the energy

$$E_{\text{CI}} = E_0 + \sum_u \mathsf{H}_{0\mu} \mathsf{C}_\mu^{\text{CI}} \tag{64}$$

and defining the PNO–CI pair correlation energy

$$\varepsilon_\mu^{\text{CI}} = \mathsf{H}_{0\mu} \mathsf{C}_\mu^{\text{CI}}. \tag{65}$$

Equation (62) then becomes

$$\begin{pmatrix} E_0 & \mathsf{H}_{0\mu} \\ \mathsf{H}_{\mu 0} & \mathsf{H}_{\mu\mu} \end{pmatrix} \begin{pmatrix} 1 \\ \mathsf{C}_\mu^{\text{CI}} \end{pmatrix} - (E_0 + \varepsilon_\mu^{\text{CI}}) \begin{pmatrix} 1 \\ \mathsf{C}_\mu^{\text{CI}} \end{pmatrix} = - \sum_{\nu \neq \mu} \begin{pmatrix} 0 \\ \mathsf{H}_{\mu\nu} \mathsf{C}_\nu^{\text{CI}} - \varepsilon_\nu^{\text{CI}} \mathsf{C}_\mu^{\text{CI}} \end{pmatrix}. \tag{66}$$

A new set of equations is then defined in which the term summing the products of the ε_ν and C_μ^{CI} is omitted:

$$\begin{pmatrix} E_0 & \mathsf{H}_{0\mu} \\ \mathsf{H}_{\mu 0} & \mathsf{H}_{\mu\mu} \end{pmatrix} \begin{pmatrix} 1 \\ \mathsf{C}_\mu^{\mathrm{CEPA}} \end{pmatrix} - (E_0 + \varepsilon_\mu^{\mathrm{CEPA}}) \begin{pmatrix} 1 \\ \mathsf{C}_\mu^{\mathrm{CEPA}} \end{pmatrix} = -\sum_{\nu \neq \mu} \begin{pmatrix} 0 \\ \mathsf{H}_{\mu\nu} \mathsf{C}_\nu^{\mathrm{CEPA}} \end{pmatrix}. \tag{67}$$

These are the CEPA equations and they lead to a total energy of the form

$$E_{\mathrm{CEPA}} = E_0 + \sum_\mu \varepsilon_\mu^{\mathrm{CEPA}}, \tag{68}$$

which, in spite of being nonvariational, has been shown to provide significantly improved results for many molecules [37, 38]. Equation (68) has the correct form for noninteracting spaces but contains the interactions between the pairs by virtue of the presence of the $H_{\mu\nu}$ in Eq. (67). The physical distinction between the IEPA and CEPA is that whereas the former treats each electron pair in the *Hartree–Fock* field of the other electrons, the latter treats each pair in the fields of the other *correlated* electrons.

A related method, termed the theory of self-consistent electron pairs (SCEP), has been published recently [39]. It combines the features of the PNO–CI approach, first-order perturbation theory, and the direct determination of the correlated wave function from the basis function matrix elements (Section B) without the construction of a Hamiltonian matrix. Since no configuration list is required, the use of large basis sets is not prohibitive and, in fact, the dependence on basis set size is nearly proportional to that of a corresponding SCF procedure multiplied by the number of pairs. This is, for N basis functions, roughly an N^3 dependence, a significant improvement over the N^5 dependence inherent in conventional configuration-mixing methods, where integral transformations are required. The SCEP method provides orbitals that are optimum in the space of single and double replacements relative to suitably chosen reference configurations and gives variationally additive pair correlation energies. Calculations on the molecules H_2, LiH, BeH^+, BH, Be_2, CH_2, H_2O, H_2CO, HCCH [39b] and Be_4 [39c] indicate that the SCEP method may be one of the more promising approaches for obtaining correlated wave functions.

REFERENCES

1. (a) I. N. Levine, "Quantum Chemistry." Allyn and Bacon, Boston, Massachusetts, 1970 (two volumes).
(b) R. McWeeny and B. T. Sutcliffe, "Methods of Molecular Quantum Mechanics." Academic Press, New York, 1969.
(c) F. L. Pilar, "Elementary Quantum Chemistry." McGraw-Hill, New York, 1968.

(d) J. C. Slater, "Quantum Theory of Molecules and Solids," Vols. I and II. McGraw-Hill, New York, 1963.
(e) D. B. Cook, "Ab-Initio Valence Calculations in Chemistry." Wiley, New York, 1974.

2. (a) H. F. Schaefer, III, "The Electronic Structure of Atoms and Molecules." Addison-Wesley, Reading, Massachusetts, 1972.
(b) H. F. Schaefer, III, Molecular electronic structure theory: 1972–1975, *Annu. Rev. Phys. Chem.* **27**, 261 (1976).
(c) H. F. Schaefer, III, ed., "Modern Theoretical Chemistry," Vols. 3 and 4. Plenum, New York, 1977.
(d) A. C. Hurley, "Introduction to the Electron Theory of Small Molecules," Academic Press, London, 1976; "Electron Correlation in Small Molecules," Academic Press, London, 1976.
3. G. Herzberg, "Molecular Spectra and Molecular Structure," Volume I, Diatomic Molecules, 2nd ed. Van Nostrand-Reinhold, Princeton, New Jersey, 1950.
4. E. A. McCullough, *Chem. Phys. Lett.* **24**, 55 (1974).
5. T. L. Gilbert and P. J. Bertoncini, *J. Chem. Phys.* **61**, 3026 (1974); T. L. Gilbert, *ibid.* **62**, 1289 (1975).
6. J. C. Slater, *Phys. Rev.* **36**, 57 (1930); B. J. Ransil, *Rev. Mod. Phys.* **32**, 245 (1960).
7. S. F. Boys, *Proc. Roy. Soc. London Ser. A* **201**, 125 (1950); I. Shavitt, *Methods Comput. Phys.* **2**, 1 (1963).
8. (a) J. L. Whitten, *J. Chem. Phys.* **44**, 359 (1966).
(b) T. H. Dunning, Jr., *ibid*, **53**. 2823 (1970).
9. E. Clementi, Tables of Atomic Functions, *Suppl. IBM J. Res. Dev.* **9**, 2 (1965); S. Huzinaga, Approximate Atomic Functions, Div. Theoret. Chem. Rep. Univ. of Alberta Canada, 1971. See also F. R. Burden and R. M. Wilson, *Adv. Phys.* **21**, 825 (1972).
10. P. E. Cade, K. D. Sales, and A. C. Wahl, *J. Chem. Phys.* **44**, 1973 (1966); T. H. Dunning, Jr., *ibid.* **55**, 3958 (1971).
11. H. Preuss, *Int. J. Quantum Chem.* **2**, 651 (1968); S. Rothenberg and H. F. Schaefer, III, *J. Chem. Phys.* **54**, 2764 (1971).
12. C. C. J. Roothaan, *Rev. Mod. Phys.* **23**, 69 (1951).
13. T. A. Koopmans, *Physica* **1**, 104 (1933).
14. (a) C. C. J. Roothaan, *Rev. Mod. Phys.* **33**, 179 (1960).
(b) S. Huzinaga, *Phys. Rev.* **120**, 866 (1960); **122**, 131 (1961); *J. Chem. Phys.* **51**, 3971 (1969).
(c) C. C. J. Roothaan and P. S. Bagus, *Methods Comput. Phys.* **2**, 47 (1963).
(d) F. W. Birss and S. Fraga, *J. Chem. Phys.* **38**, 2552 (1963); **40**, 3203 (1964).
(e) W. J. Hunt, T. H. Dunning Jr., and W. A. Goddard, III, *Chem. Phys. Lett.* **3**, 606 (1969); **4**, 231 (1969); **6**, 147 (1970).
(f) D. Peters, *J. Chem. Phys.* **57**, 4751 (1972).
(g) P. S. Bagus, Res. Rep. RJ1077, IBM Corp., San Jose, California.
(h) E. R. Davidson and L. Stenkamp, *Int. J. Quantum. Chem.* **105**, 21 (1976).
(i) D. R. Yarkony, H. F. Schaefer, III, and C. F. Bender, *J. Chem. Phys.* **64**, 981 (1976).
15. R. K. Nesbet, *Proc. Roy. Soc. London Sect. A* **230**, 312 (1955).
16. P.-O. Löwdin, *Adv. Chem. Phys.* **2**, 207 (1159).
17. Z. Gershgorn and I. Shavitt, *Int. J. Quantum Chem.* **1S**, 403 (1967).
18. (a) A. D. McLean and B. Liu, *J. Chem. Phys.* **58**, 1066 (1973).
(b) P. S. Bagus, B. Liu, A. D. McLean, and M. Yoshimine, "Wave Mechanics: The Fifty Years" (W. C. Price, S. S. Chissick, and T. Ravensdale, eds.). Butterworth, London, 1973.
(c) I. Shavitt, "Modern Theoretical Chemistry" (H. F. Schaefer, III, ed.), Vol. 3. Plenum, New York, 1977.

19. P.-O. Löwdin, *Phys. Rev.* **97**, 1474 (1955).
20. E. R. Davidson, *Adv. Quantum Chem.* **6**, 235 (1972).
21. B. Roos, *Chem. Phys. Lett.* **15**, 153 (1972); R. F. Hausman, Jr., S. D. Bloom, and C. F. Bender, *ibid.* **32**, 483 (1975).
22. Z. Gershgorn and I. Shavitt, *Int. J. Quantum Chem.* **2**, 751 (1968).
23. (a) A. Veillard and E. Clementi, *Theor. Chim. Acta* **1**, 133 (1967).
    (b) G. Das and A. C. Wahl, *J. Chem. Phys.* **47**, 2934 (1967).
    (c) T. L. Gilbert, *Phys. Rev.* **A6**, 580 (1972).
    (d) J. Hinze, *J. Chem. Phys.* **59**, 6424 (1973).
    (e) A. C. Wahl and G. Das, *in* "Modern Theoretical Chemistry" (H. F. Schaefer, III, ed.), Vol. 3. Plenum, New York, 1977.
24. (a) P.-O. Löwdin, *Phys. Rev.* **97**, 1474, 1490, 1509 (1955).
    (b) P.-O. Löwdin, *Adv. Chem. Phys.* **2**, 207 (1959).
    (c) P.-O. Löwdin, *Rev. Mod. Phys.* **32**, 328 (1960).
25. K. Ruedenberg, *Rev. Mod. Phys.* **34**, 326 (1962).
26. E. R. Davidson, "Reduced Density Matrices in Quantum Chemistry." Academic Press, New York, 1976.
27. I. Shavitt, B. J. Rosenberg, and S. Palalikit, *Int. J. Quantum. Chem.* **S10**, 33 (1976).
28. C. F. Bender and E. R. Davidson, *J. Phys. Chem.* **70**, 2675 (1966).
29. P. J. Hay, *J. Chem. Phys.* **59**, 2468 (1973); A. K. Siu and E. F. Hayes, *ibid.* **61**, 37 (1974).
30. W. Kutzelnigg, *Top. Current Chem.* **41**, 31 (1973).
31. R. K. Nesbet, *Adv. Chem. Phys.* **9**, 321 (1965); **14**, 1 (1969).
32. O. Sinanoglŭ, *Adv. Chem. Phys.* **6**, 315 (1964); **14**, 337 (1969).
33. A. C. Hurley, J. Lennard-Jones, and J. A. Pople, *Proc. Roy. Soc. London* **A220**, 446 (1953).
34. W. Kutzelnigg, *J. Chem. Phys.* **40**, 3640 (1964).
35. H. J. Silverstone and O. Sinanoglŭ, *J. Chem. Phys.* **44**, 1899, 3608 (1966).
36. C. Edmiston and M. Krauss, *J. Chem. Phys.* **45**, 1833 (1966).
37. W. Meyer, *J. Chem. Phys.* **58**, 1017 (1973).
38. R. Ahlrichs, H. Lischka, V. Straemmler, and W. Kutzelnigg, *J. Chem. Phys.* **62**, 1225 (1975).
39. (a) W. Meyer, *J. Chem. Phys.* **64**, 2901 (1976).
    (b) C. E. Dykstra, H. F. Schaefer, III, and W. Meyer, *ibid.* **65**, 2740 (1976).
    (c) C. E. Dykstra, H. F. Schaefer, III, and W. Meyer, *ibid.* **65**, 5141 (1976).

CHAPTER II

ONE-ELECTRON MOs AS PROTOTYPES

A. H_2^+ MOs IN ELLIPTICAL COORDINATES

Experimentally, the H_2^+ ion is well known in mass spectrometry, but the only other experimental data are on hyperfine structure in its vibration–rotation levels [1]. Hence for our understanding of this molecule, we have to rely almost entirely on theoretical calculations. When $\mathscr{H}$ in the time-independent Schrödinger equation $\mathscr{H}\Psi = E\Psi$ for H_2^+, with the nuclei held fixed at a constant distance R, is transformed to elliptical coordinates, the equation is separable; that is, it can be solved in the exactly factored form $\Psi = \Lambda(\lambda)\, M(\mu)\, \Phi(\phi)$, where λ and μ are the usual confocal elliptic coordinates and ϕ is the azimuthal angle [2]. (Do not confuse λ and Λ as used here with the like-named quantum numbers.) Here $\lambda = (r_a + r_b)/2R$ and $\mu = (r_a - r_b)/2R$, where r_a and r_b are the distances of the electron from the two nuclei (see Fig. 1). For this (and other) 1-electron molecules or ions, each Ψ is itself an MO. The H_2^+ MOs are prototypes of those of homopolar diatomic polyelectron molecules, where the Ψ's, though expressible using MOs, are not themselves MOs.

The forms of Λ and M can be obtained as series expansions, while $\Phi = \exp(im\phi)$ or $\exp(-im\phi)$. Our main interest will be in the nodal properties of Λ and M. The nodes are respectively ellipsoids and hyperboloids of revo-

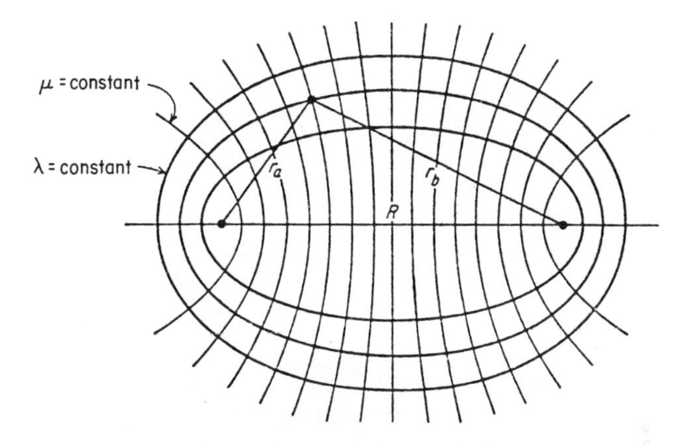

FIG. 1 Spheroidal coordinates for the two-center problem.

lution. As E increases, the numbers of ellipsoidal, hyperboloidal, or ϕ nodes increases. The number of nodes in each coordinate defines a corresponding quantum number (n_λ, n_μ, m). The principal quantum number n is equal to $1+n_\lambda+n_\mu+|m|$.

The behavior of the nodal surfaces as $R\to 0$ (*united atom*) and $R\to\infty$ are of especial interest. At $R=0$, $\lambda\to r$ and $\mu\to\theta$: the nodal surfaces become respectively spheres and cones; ϕ is unchanged in meaning. At $R=\infty$, for the portions of Ψ near the two nuclei, the nodal surfaces of λ and μ both become paraboloids of revolution with their foci at the nuclei; the portions of Λ and M *not* near the nuclei fade to nothingness as $R\to\infty$.

The lowest energy solution $1\sigma_g$ has no nodes, and has the full $D_{\infty h}$ symmetry of the nuclear potential function. It is of course concentrated mainly near and between the two nuclei. As $R\to 0$, it becomes 1s of the united atom He^+. As $R\to\infty$, the part between the nuclei fades away, and what is left is seen, when normalized, to take precisely the form $(1s_a+1s_b)/2^{1/2}$, where $1s_a$ and $1s_b$ are 1s AOs of H atoms a and b. At smaller R values, normalization requires that account be taken of the overlap between χ_a and χ_b, and the normalized form becomes $(1s_a+1s_b)/2^{1/2}(1+S)^{1/2}$, where S is the overlap integral $\int \chi_a \chi_b \, dV$. In the following, the simple form $1s_a+1s_b$, ignoring normalization, will be called $\sigma_g 1s$, meaning σ_g derived from 1s. Here in H_2^+, 1s is an H atom AO, but later we shall use it

(a) to refer to a 1s AO of any atom, or alternatively
(b) to refer to a 1s STF, a major building block for a 1s AO.

In case (a) we refer to $\sigma_g 1s$ as an LCAO (linear combination of AOs) expression, in case (b) as an LCSTF expression. Symbols analogous to $\sigma_g 1s$ will be used for other MOs or building blocks of MOs (see below).

Actually at very large R for H_2^+ we should use solutions of the time-dependent Schrödinger equation, of the form $1s_a \cos \nu t + 1s_b \sin \nu t$, where $h\nu$ is the energy difference ΔE between the two MOs $1\sigma_g$ and $1\sigma_u$, which vanishes as $R \to \infty$. At moderate R values, the simple LCAO form $\sigma_g 1s$ of $1\sigma_g$ is no longer exact, but more general LCSTF expressions [see Eq. (1) below] or else the exact form $\Lambda M \Phi$ must be used.

The excited MOs of H_2^+ may be approximated by first considering their LCAO forms at $R = \infty$. The lowest-energy excited MO $1\sigma_u$ is then of the LCAO form $(1s_a - 1s_b)/\sqrt{2}$, or $\sigma_u 1s$ with a nodal plane midway between the two nuclei. On decreasing R, the midplane is seen to be the hyperboloidal surface $\mu = 0$, with the quantum number $n_\mu = 1$. As $R \to 0$, it becomes the one nodal surface, with $n_\theta = 1$, of a $2p\sigma$ unite-atom AO.

Next we consider the LCAO forms derivable from 2-quantum H atom AOs. For the two atoms, we now need Stark-effect AOs. (As $R \to \infty$, H_2^+ consists [1] of an H atom in the uniform electric field of an infinitely distant H^+ nucleus.) In the Stark effect, these AOs when expressed in terms of ordinary spherical coordinate AOs include the normalized digonal hybrids $2di^{in} \equiv (2s + 2p\sigma)/2^{1/2}$ and $2di^{out} \equiv (2s - 2p\sigma)/2^{1/2}$, where the superscript "in" (inward-pointing) refers to a paraboloidal AO on nucleus a with its vertex pointing toward b, or on b pointing toward a, and the superscript "out" (outward-pointing) refers to a paraboloidal AO with its vertex pointing away from the other nucleus. There is also the twofold degenerate AO $2p\pi$. (Of course $2di^{in}$, $2di^{out}$, and $2p\pi$ are all degenerate in the free H atom in the absence of a field, but in the field of a proton, as in H_2^+, their energies begin to diverge.) In terms of H atom paraboloidal quantum numbers, $n_{\lambda'} = 1$, $n_{\mu'} = m = 0$ for $2di^{out}$, $n_{\lambda'} = 0$, $n_{\mu'} = 1$, $m = 0$ for $2di^{in}$, and $n_{\lambda'} = n_{\mu'} = 0$, $m = \pm 1$ for $2p\pi$.

The corresponding normalized MOs, in order of increasing energy at large R, with their LCAO forms as $R \to \infty$, are:

$2\sigma_g \approx (2di_a^{in} + 2di_b^{in})/2^{1/2}$ with $n_\lambda = 0$, $n_\mu = 2$, $m = 0$, becoming $3d\sigma$ as $R \to 0$;

$3\sigma_g \approx (2di_a^{out} + 2di_b^{out})/2^{1/2}$ with $n_\lambda = 1$, $n_\mu = m = 0$, becoming 2s as $R \to 0$; its potential curve crosses that of $n_\lambda = 0$, $n_\mu = 2$, $m = 0$ at smaller R, so that it is now $2\sigma_g$ and $n_\lambda = 0$, $n_\mu = 2$, $m = 0$ is $3\sigma_g$;

$1\pi_u \approx (2p\pi_a + 2p\pi_b)/2^{1/2}$ with $n_\lambda = n_\mu = 0$, $|m| = 1$, becoming $2p\pi$ as $R \to 0$;

$1\pi_g \approx (2p\pi_a - 2p\pi_b)/2^{1/2}$ with $n_\lambda = 0$, $n_\mu = |m| = 1$, becoming $3d\pi$ as $R \to 0$;

$2\sigma_u \approx (2di_a^{out} - 2di_b^{out})/2^{1/2}$ with $n_\lambda = n_\mu = 1$, $m = 0$, becoming $3p\sigma$ as $R \to 0$;

$3\sigma_u \approx (2di_a^{in} - 2di_b^{in})/2^{1/2}$ with $n_\lambda = 0$, $n_\mu = 3$, $m = 0$, becoming $4f\sigma$ as $R \to 0$.

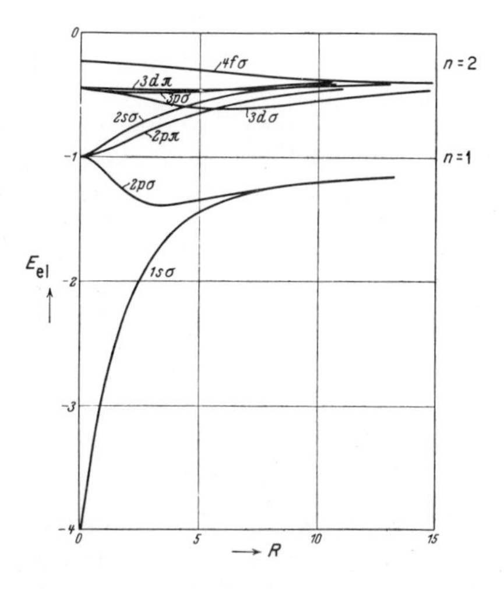

FIG. 2 Electronic energy of H_2^+ in Rydberg units (1 Ry.u. = 0.5 a.u.) as a function of R (in a.u.). The positions of the $n = 1$ and $n = 2$ levels of the H atom are indicated at the right. [Adapted from E. Teller, *Z. Phys.* **61**, 458 (1930).]

These MOs at moderate R values may be denoted briefly as $\sigma_g 2di^{in}$, $\sigma_g 2di^{out}$, $\pi_u 2p$, $\pi_g 2p$, $\sigma_u 2di^{out}$, and $\sigma_u 2di^{in}$. The correlation diagram Fig. 2 shows how the electronic energies of the various MOs just described change as we go from H plus H^+ at $R = \infty$ to He^+ at $R = 0$.

In all homopolar diatomic molecules there is a tendency for the MOs to resemble those of H_2^+, but the LCAO approximations to the MOs differ from those of H_2^+ because only in the H atom are AOs of equal n degenerate so that the use for example of digonal hybrids in the LCAO expressions for the σ MOs related to $n = 2$ H atom AOs is required. Thus $2\sigma_g$ and $3\sigma_g$ in LCAO approximations for most first-row homopolar diatomics are more nearly $\sigma_g 2s$ and $\sigma_g 2p$ respectively, and $2\sigma_u$ and $3\sigma_u$ more nearly $\sigma_u 2s$ and $\sigma_u 2p$. A very considerable amount of 2s, $2p\sigma$ hybridization does persist, however, although much less complete than digonal.

The Stark-effect H atom AOs needed to describe the large-R LCAO forms of the higher energy MOs of H_2^+ involve increasing hybridization in terms of spherical-coordinate AOs. Thus for $n = 3$, three σ hybrids (of 3s, $3p\sigma$, and $3d\sigma$) are required, and two π hybrids (of $3p\pi$ and $3d\pi$), but $3d\delta$ is unhybridized. We shall not attempt to discuss these higher energy MOs of H_2^+ further.

The relations of the MOs of, for example, H_2^+ to the AOs of their LCAO approximations involve "*promotion*," which means the creation of new nodal surfaces when a molecule is formed. The promotions involved in H_2^+ in the $n = 1$ and 2 H atom cases already discussed can be summarized as shown

TABLE 1

	$n_{\lambda'}$	$n_{\mu'}$	m	n_λ	n_μ	m	
$1\sigma_g$	0	0	0	0	0	0	Unpromoted
$1\sigma_\mu$	0	0	0	0	1	0	Promoted (new node in μ)
$2\sigma_g$	0	1	0	0	2	0	Promoted ($n_\mu = 2n_{\mu'}$)
$3\sigma_g$	1	0	0	1	0	0	Unpromoted
$1\pi_u$	0	0	± 1	0	0	± 1	Unpromoted
$2\pi_g$	0	0	± 1	0	1	± 1	Promoted (new node in μ)
$2\sigma_u$	1	0	0	1	1	0	Promoted (new node in μ)
$3\sigma_u$	0	1	0	0	3	0	Promoted ($n_\mu = 2n_{\mu'}+1$)

in Table 1. In the *additive* LCAO MOs (of the form $\chi_a+\chi_b$), promotion occurs only in $n_{\mu'}$, in which case $n_\mu = 2n_{\mu'}$. In the subtractive MOs (of the form $\chi_a-\chi_b$), there is an additional promotion in $n_{\mu'}$ by the creation of a nodal midplane. Valence-shell additive MOs (here $1\sigma_g$) are *bonding*, valence-shell subtractive MOs (here $1\sigma_u$) are *antibonding*. Excited-shell MOs are nearly nonbonding.

B. LCAO AND LCSTF EXPANSIONS; BASIS SETS; LCMAOs

At this point we enter on a discussion much of which is applicable to diatomic molecules in general. At small and moderate R values, the simple LCAO forms of MOs which are valid as $R \to \infty$ are no longer adequate. For homopolar diatomics, however, they can be generalized to expansions of LCSTFs such as σ_g 1s, π_u2p, etc.:

$$\phi_i = \sum_p c_{ip}(\chi_{pa} \pm \chi_{pb}). \tag{1a}$$

Here each χ_p is an STF. A corresponding equation for heteropolar diatomics is

$$\phi_i = \sum_{m,n} (c_{ima}\chi_{ma} + c_{inb}\chi_{nb}). \tag{1b}$$

One can alternatively use expressions in terms of elliptical coordinates, but the expansions in terms of STFs are usually more convenient for diatomics in general.

What is meant by an STF can be seen by writing detailed expressions for the forms of the AOs of the H atom or of the SCF AOs of any polyelectron atom. These take the form

$$\chi_{nlm} = \sum_{j=l}^{n-1} \sum_k c_{jk} r^j \exp(-\zeta_{jk} r) Y_{lm}(\theta, \phi), \tag{2}$$

where the Y_{lm}'s are spherical harmonics, l is the azimuthal and n the principal quantum number of the AO, and ζ_{jk} is an "orbital exponent." Here r is in atomic units ($r = 1$ is a_0, the Bohr radius of the H atom in its ground state). ***Each term*** in Eq. (2), after normalization, is an STF. For example, the following are normalized STFs:

$$1s^{(\zeta)} = (\zeta^3/\pi)^{1/2} e^{-\zeta r}, \qquad 2s^{(\zeta)} = (\zeta^3/3\pi)^{1/2} r e^{-\zeta r},$$

$$2p\sigma^{(\zeta)} = (\zeta^3/\pi)^{1/2} r e^{-\zeta r} \cos\theta, \qquad 2p\pi^{(\zeta)} = (\zeta^3/\pi)^{1/2} r e^{-\zeta r} \sin\theta \, e^{\pm i\phi}.$$

For the free H atom, $\sum_k$ reduces to a single term, say $k = 1$, and ζ_{jk} is given by $1/(j+1)$ precisely. For atoms with more than one electron, $\sum_k$ is an infinite series if the AO is to be represented precisely, but in practice it is usually approximated by using not more than four terms. In SCF AO calculations, STFs with $j > n-1$ in Eq. (2) have sometimes been included. When STFs are used to build up an accurate MO [Eqs. (1)] additional STFs (polarization functions) must in general be included with l values greater than those in the AOs of the atoms from which the molecule is formed.

In machine computations, the collection of STFs used is called a *basis set*. A "minimal set," with $\sum_k$ in Eq. (2) reduced to a single term, usually is composed of as many STFs as there are AOs in the simple LCAO forms which are valid as $R \to \infty$: namely, on a first-row atom, one 1s with $j = 0$, one 2s, and in general one 2pσ and one 2pπ all with $j = 1$. The use of a minimal basis set results in rather crude approximations to the exact MOs. If, however, such a set is used, it should if possible be *optimized*; that is, all ζ values should be chosen by a trial and error process so as to minimize the computed energy.

With an *extended* (more extensive) basis set of STFs, the choice of orbital exponents becomes less critical. Experience has shown, for molecules in general, however, that a good choice of STFs in Eq. (1) is one using ζ values which have been optimized for relevant *atomic* SCF AOs [Eq. (2)] and then adding polarization functions.

Another procedure which minimizes the time-consuming problem of optimizing ζ's is to use an *even-tempered basis set*, as proposed by Ruedenberg [3]. Here for each *species* of STF (s, pσ, pπ, etc.), an STF with a suitable ζ, say α, is chosen together with additional STFs of the same species whose ζ's differ from this ζ by integral powers of a constant factor β (somewhere in the neighborhood of 1.5 or 2), thus constituting a set of ζ's which form a geometrical progression: $\zeta_i = \alpha_i \beta_i^{\,k}$; $\alpha_i > 0$, $b_i > 1$, $k = 1, 2, \ldots, m$. With an even-tempered set, it is sufficient to work with the simplest forms of STFs, those with the smallest j values in Eq. (2) for each species, namely 1s, 2pσ, 2pπ, 3dσ, 3dπ, and so on. Although it is mentioned here under H_2^+, the main use for even-tempered sets is for larger molecules (see Section V.A for the N_2 MOs as an example).

TABLE 2

Approximate Forms of H_2^+ MOs[a]

MO	R (a.u.)	Coefficients and (in parentheses) ζ values of STFs			Total energy (a.u.)
		σ_g 1s	σ_g 2p	σ_g 3d	
$1\sigma_g$	0.5	0.476 (1.744)	0.193 (1.439)	0.031 (1.912)	0.2653
	2.0	0.524 (1.244)	0.116 (1.152)	0.041 (1.333)	−0.6020
	4.0	0.615 (1.047)	0.078 (0.975)	0.026 (1.063)	−0.5458
	10.0	0.705 (0.999)	0.017 (0.790)	0.004 (0.788)	−0.5006
		σ_u 1s	σ_u 2p	σ_u 3d	
$1\sigma_u$	0.5	0.791 (0.641)	−0.437 (1.054)	−0.007 (2.348)	1.4831
	2.0	1.012 (0.951)	−0.073 (1.253)	−0.023 (1.522)	−0.1670
	4.0	0.816 (0.984)	0.042 (0.791)	0.006 (0.757)	−0.4455
	10.0	0.708 (1.000)	0.016 (0.797)	0.002 (0.834)	−0.4999
		π_u 2p	π_u 3d		
$1\pi_u$	0.5	0.501 (0.982)	0.020 (1.169)		1.5077
	2.0	0.512 (0.851)	0.065 (1.057)		0.0717
	4.0	0.532 (0.709)	0.106 (0.864)		−0.0998
	10.0	0.614 (0.529)	0.137 (0.559)		−0.1317

[a] The coefficients are the c_{ip}'s of Eq. (1a).

To illustrate the use of Eq. (1), in Table 2 there are for H_2^+ some expansions for the first three MOs, at each of several R values (R in atomic units); for exact representations, more terms would of course be needed.

For the *additive* MOs $1\sigma_g$ and $1\pi_u$ of H_2^+, the main term in the LCSTF expansion is qualitatively the same as the LCAO form valid as $R \to \infty$, but its optimal ζ value increases toward the united-atom value as R decreases. At $R = \infty$, Eq. (2) reduces to a single STF with $\zeta = 1$ for 1s and 0.5 for $2p\pi$, while at $R = 0$, there is again a single term with $\zeta = 2$ for 1s and 1 for 2p. The subsidiary terms, which may be looked on as representing a type of hybridization, or of polarization, become important only at intermediate R values. As $R \to 0$, the main term $1s_a + 1s_b$ or $2p\pi_a + 2p\pi_b$ approaches simply 1s or $2p\pi$ of the united atom He^+, of $Z = 2$.

For the *subtractive* MO $1\sigma_u$, the main term $1s - 1s$ at larger R values gives way as $R \to 0$ to the term $2p\sigma + 2p\sigma$, and finally disappears completely at $R = 0$. (One can show that the limit as $R \to 0$ of the form $1s - 1s$ is $\exp(-\zeta r/a_0) \cos\theta$, which might be called $1p\sigma$, but which is unacceptable as a solution of the Schrödinger equation.) This type of disappearance, which is characteristic of all subtractive MOs, takes care of an otherwise paradoxical fact: namely, if one considers the complete infinite set of all AOs and MOs, the basis set of AOs at $R = \infty$ is twice as numerous as at $R = 0$, since at $R = \infty$ there is a complete infinite set of like AOs for each of two atoms, whereas at $R = 0$ there is only one such set. The *redundancy* which exists at intermediate and large R values is extinguished as $R \to 0$ (see Fig. 2). Thus one sees that while the LCAO approach from $R = \infty$ suggests that there are twice as many MOs for H_2^+ as for an H atom, in fact the number is no greater. For polyelectron molecules the corresponding paradox becomes much more severe, but can be resolved in similar fashion.

An instructive alternative to the viewpoint of Eq. (1) at intermediate R values is to maintain the simple LCAO idea by thinking of $\sum_m c_{ima} \chi_{ma}$ and $\sum_m c_{imb} \chi_{mb}$ each as a single *modified* AO, or MAO, modified, as compared with a free-atom AO, by scaling (changes in ζ value), symmetrical distortions [readjustment of coefficients in Eq. (1a)], and polarization or hybridization (introduction of additional terms). Then all MOs are simple LCMAOs:

$$\phi_i = (\chi_a{}^{\mathrm{M}} + \chi_b{}^{\mathrm{M}})/2^{1/2}(1 \pm S^{\mathrm{M}})^{1/2}, \tag{3}$$

where the χ^{M}'s are MAOs. Here S^{M} is the (modified) overlap integral $\int \chi_a{}^{\mathrm{M}} \chi_b{}^{M} \, dv$. Equation (3) is readily generalized for the heteropolar case. Reference should be made to Ref. 4a for further details on MAOs.

Instead of expansion in terms of STFs which are related to *separate-atom* AOs (SAOs), it is also possible to use expansions in terms of single-center functions centered at the molecular midpoint. These may be regarded as STFs related to united-atom AOs (UAOs). Experience has shown, however, that such expansions converge only slowly, and they are no longer much used. Nevertheless, if enough terms with higher l values are used, accurate results can be obtained (within 1 or 2 kcal/mole) by extrapolation, as Hayes and Parr have shown in the case of H_2^+ [4b]. An expansion using *both* types of STFs sometimes has advantages, as will be illustrated in later chapters.

Because of the complexity of 3- and 4-center electron repulsion integrals involving STFs, it is often more convenient for polyatomic molecules to use linear combinations of Gaussian-type functions, or Gaussian-lobe functions (Section I.A).

C. SPECTROSCOPIC TRANSITION PROBABILITIES

For any atom or molecule, the intensity of a spectroscopic transition between two electronic states m and n can be expressed in terms of the components μ_{mn}^q of the transition dipole moment, as follows (in atomic units, after putting $e = 1$ in the dipole moment operator $e \sum_i q^i$):

$$\mu_{mn}^q = \int \Psi_m{}^* \left(\sum_i q_i \right) \Psi_n \, dv, \tag{4}$$

where q may be x, y, or z (or $x \pm iy$ and z) and the summation $\sum_i$ is over all electrons. In general, μ_{mn} is a vector with x, y, and z components, or some of these. Two alternatives to the use of the operator $\sum_i q_i$ in Eq. (4) are the employment of velocity or acceleration operators [5]. If Ψ is exact, identical results are obtained; otherwise there are discrepancies.

Consider for example the transition $1\sigma_u \leftrightarrow 1\sigma_g$ in H_2^+ at an internuclear separation R. Here Ψ_m is $1\sigma_g$ and Ψ_n is $1\sigma_u$, and $\sum q_i$ is just z if the symmetry axis of the ion is chosen as z axis ($\mu_{mn}^x = \mu_{mn}^y = 0$ because of the forms of $1\sigma_g$ and $1\sigma_u$). Using the LCMAO expressions of Eq. (3), we obtain

$$\mu = \mu^z$$
$$= \int [1s_a{}^M + 1s_b{}^M)/2^{1/2}(1+S^M)^{1/2}]\, z[(1s_a^{M'} - 1s_b^{M'})/2^{1/2}(1-S^{M'})^{1/2}]\, dv. \tag{5a}$$

Neglecting the fact that $1s^{M'}$ and $S^{M'}$ differ considerably from $1s^M$ and S^M, we obtain as a rather rough approximation

$$\mu \approx \tfrac{1}{2}(z_a - z_b)/(1-S^2)^{1/2} = \tfrac{1}{2}R/(1-S^2)^{1/2}, \tag{5b}$$

since $\int (1s_a{}^M)^2\, dv = \int (1s_b{}^M)^2\, dv = 1$.

In *absorption*, a good measure of the intensity is given by the oscillator strength f, which is proportional[1] to the product of $(\mu^q)^2$ and ν (the transition energy) summed over all q which contribute. (In general, when m and/or n are degenerate states, there may be a μ_{mn} for more than one q.) The oscillator strength is defined as

$$f_{mn} = 1.085 \times 10^{-5}\, \nu_{mn} \sum_q (\mu_{mn}^q)^2 \tag{6a}$$

if ν_{mn} is in cm^{-1}. Although Eq. (5b) shows that μ increases at first approximately in proportion to R, the exponential decrease in ν toward zero as $R \to \infty$ causes f to reach a maximum and then go to zero as $R \to \infty$.

For the *emission* intensity I_{mn} (in ergs/second/molecule), we have

$$I_{mn} = 1.436 \times 10^{-21}\, \nu_{mn}^4 \sum_q (\mu_{mn}^q)^2. \tag{6b}$$

From the exact wave functions for H_2^+, Bates and collaborators have made exact computations of oscillator strengths for various absorption transitions of H_2^+ [7]. They have also computed some exact cross sections for photoionization [8].

D. HeH^{2+} MOs

As a (not very good) prototype for *heteropolar* diatomic molecules we may consider HeH^{2+}. The Schrödinger equation for one-electron molecules

[1] See e.g. Mulliken [6] but with an improved numerical factor.

orions, like that of H_2^+, is separable in elliptical coordinates, but the MOs are now unsymmetrical.[2] Using a minimal STF basis set, we obtain

$$\phi_i \approx c_{ia}\chi(\mathrm{He}^+) + c_{ib}\chi(\mathrm{H}), \tag{7}$$

where c_{ia} and c_{ib} are unequal. As $R \to \infty$, one of c_{ia} and c_{ib} vanishes. As examples, the three lowest-energy MOs of HeH^{2+} at medium-small R values are approximately

$$\phi_{1\sigma} \approx c_{1a\sigma}\chi_{1s}(\mathrm{He}^+) + c_{1b\sigma}\chi_{1s}(\mathrm{H}), \qquad c_{1a\sigma} \gg c_{1b\sigma},$$

$$\phi_{2\sigma} \approx c_{2a\sigma}\chi_{2p\sigma}(\mathrm{He}^+) + c_{2b\sigma}\chi_{1s}(\mathrm{H}), \qquad c_{2a\sigma} > c_{2b\sigma},$$

$$\phi_{1\pi} \approx c_{1a\pi}\chi_{2p\pi}(\mathrm{He}^+) + c_{1b\pi}\chi_{2p\pi}(\mathrm{H}), \qquad c_{1a\pi} \gg c_{1b\pi}.$$

E. POPULATION ANALYSIS

1. H_2^+ Case

Writing an H_2^+ MO in the minimal-basis-set approximation $\phi = a\chi_a + b\chi_b$, the charge density ϕ^2 is

$$a^2\chi_a^{\,2} + 2ab\chi_a\chi_b + b^2\chi_b^{\,2}. \tag{8}$$

In H_2^+, $a = \pm b$, but for later reference, we write equations corresponding to the general case. On integrating over all space and using the normalization condition $\int \phi^2\,dv = 1$, Eq. (8) yields

$$a^2 + 2abS + b^2 = 1. \tag{9}$$

In H_2^+, $a = \pm b = 1/2^{1/2}(1 \pm S)^{1/2}$, and (9) becomes

$$1/[2(1 \pm S)] \pm S/(1 \pm S) + 1/[2(1 \pm S)] = 1, \tag{10}$$

which is seen to divide the total electron population (here 1) into three parts, one associated with χ_a (the *net population* on a), one associated with the overlap of χ_a and χ_b (*overlap population* n_{ab}), and the net population on b [11]. Similarly for Eq. (9) in general. This division of the total population into parts is illustrated for $1\sigma_g$ and $1\sigma_u$ of H_2^+ in Fig. 3. Note that in bonding MOs such as $1\sigma_g$, population is transferred from a and b into the overlap region (n is positive), whereas in antibonding MOs such as $1\sigma_u$, the transfer is in the opposite direction (n is negative).

[2] Exact wave functions have been obtained by Bates and Carson [9]. Wilson and Gallup [10] give graphs of the electronic energy as a function of R for $^2\Sigma$ states of HeH^{2+}, LiH^{2+}, LiH^{4+}, and He_2^{3+}.

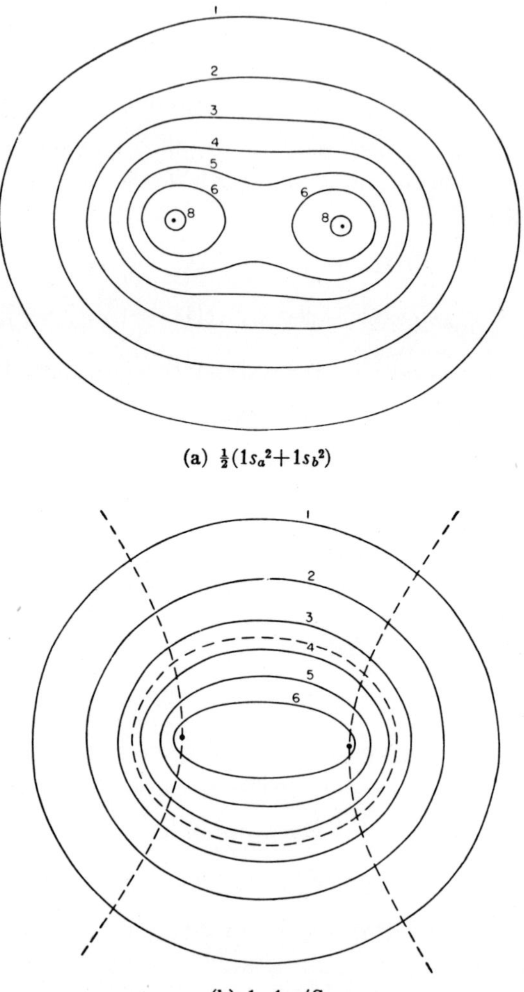

FIG. 3 Normalized population distribution functions for the electron in H_2^+: (a) mean of free-atom distributions; (b) overlap function distribution. [From R. S. Mulliken, *J. Chem. Phys.* **23**, 1833 (1955).]

If one asks for a total or *gross population* N_a or N_b on a or b, this is obviously equal to the sum of the net population plus *half* of the overlap population [11]:

$$N_a = N_b = \tfrac{1}{2}/(1 \pm S) \pm \tfrac{1}{2}S/(1 \pm S) = \tfrac{1}{2}. \tag{11}$$

Equations (8)–(11) can be generalized in an obvious way if one replaces

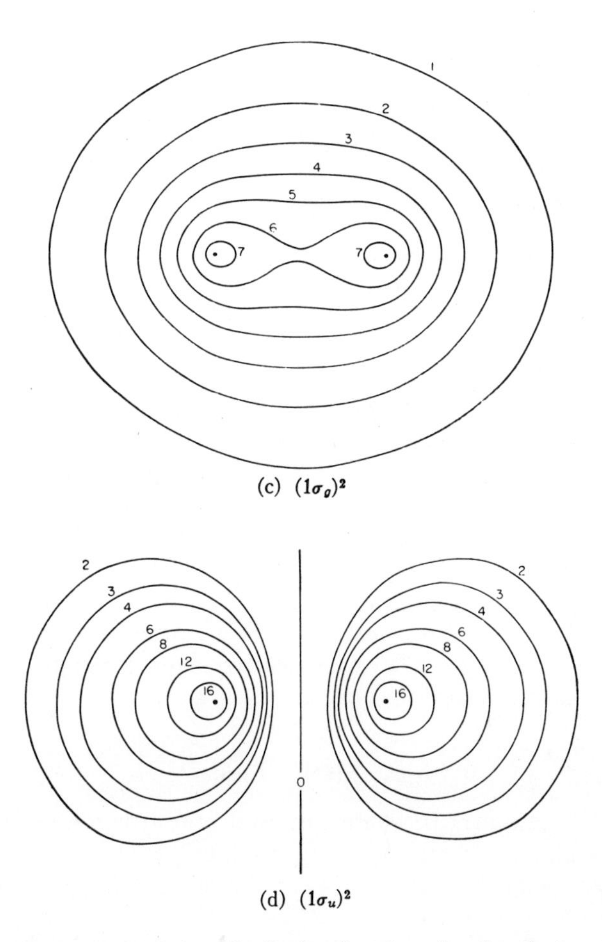

FIG. 3 Normalized population distribution functions for the electron in H_2^+: (c) distribution in ground-state MO; (d) distribution in first excited MO. [From R. S. Mulliken, *J. Chem. Phys.* **23**, 1833 (1955).]

the minimal-set ϕ's by extended LCSTF expansions as in Eq. (1). One then obtains a number of individual net and overlap population terms. These can be summed to give *total* net, overlap, or gross populations in each of the various species (s, p, d, etc., or σ, π, etc.) of STFs in Eq. (1), or combinations of these in the case of overlap populations. Alternatively, a simple analysis in terms of MAOs [cf. Eq. (3)] can be made.

To illustrate the results of population analysis numerically, let us apply it to the approximate 1σ functions given in Table 2 for $R = 2.0$ a.u. The total overlap population n is $2 \sum_{m \neq n} S_{mn} c_m c_n$ summed over all products between 1s, 2pσ, and 3dσ STFs, one on each atom, with Eq. (1) coefficients

c_{ip} as given in Table 2. The result is $n = 0.341$ for $1\sigma_g$, and $n = -1.564$ for $1\sigma_u$. The net population on each nucleus, which is obviously $\frac{1}{2}(1-n)$, is 0.330 for $1\sigma_g$ and 1.282 for $1\sigma_u$.

The gross s, p, and d populations N_s, N_p, and N_d are obtained from the net atomic populations and the overlap populations; for example, for $1\sigma_g$,

$$N_p = 2(S_{2p\sigma, 2p\sigma} c_{p\sigma}^2 + S_{1s, 2p\sigma} c_{1s} c_{2p\sigma} + S_{2p\sigma, 3d\sigma} c_{2p\sigma} c_{3d\sigma}).$$

While the meanings of these gross populations should not be taken too literally in simple cases such as H_2^+ and H_2, since they represent essentially polarizations only of AOs to form MAOs, in bigger molecules (also in H_2^+ at large R) they are often of interest as displaying substantial hybridizations. The gross populations for the MOs just discussed are: $N_s = 0.890$, $N_p = 0.092$, $N_d = 0.018$ for the $1\sigma_g$ and $N_s = 0.889$, $N_p = 0.093$ for $1\sigma_u$, $N_d = 0.018$.

The same equations (8)–(11) as for H_2^+ are applicable for any MO of a homopolar polyelectronic diatomic molecule. Here, however, the total population in any MO is not restricted to 1, but may be 2, or can be more if the MO is degenerate.

In any *additive* valence-shell MO, the overlap population is positive (with rare exceptions with hybrid AOs), while for a *subtractive* valence-shell MO it is negative. MOs with positive or negative overlap populations are respectively *bonding* and *antibonding*.

2. HeH^{2+} Case

The HeH^{2+} case may be considered as a prototype for heteropolar diatomic molecules in general. Equations (8) and (9) are now applicable. Here the overlap population is $2abS$. If we divide this equally between a and b, the gross populations are

$$N_a = a^2 + abS, \qquad N_b = b^2 + abS. \tag{12}$$

Equations (8), (9), and (12) can be generalized in an obvious manner, in the same way as for the homopolar case.

These equations are commonly used in estimating gross populations, but unless $a = b$ they are open to objections which sometimes become serious, especially for antibonding MOs. The difficulty can be seen from the following discussion.[3] Let us here consider the simple LCAO case, or more generally, we may use the MAO viewpoint of Eq. (3) and let the χ's in Eq. (8) refer to MAOs. Let $a = \cos\gamma$ and $b = \sin\gamma$. Then

$$\phi = \cos\gamma\, \chi_a + \sin\gamma\, \chi_b, \tag{13}$$

[3] Based on an unpublished communication from K. Ruedenberg, to whom we are greatly indebted for permission to reproduce the discussion and Fig. 4.

where χ_a and χ_b may be MAOs. Now if Eq. (12) is valid,

$$N_a = (\cos^2\gamma + S\sin\gamma\cos\gamma)/(1+S\sin 2\gamma)$$
$$= \tfrac{1}{2}[1+\cos 2\gamma/(1+S\sin 2\gamma)]. \quad (14)$$

An examination of the consequences of Eq. (14) shows that in certain ranges of γ and S, N_a is negative and, since $N_a+N_b=1$ for HeH^{2+}, N_b is then greater than 1. Both these results are obviously nonsense for the gross populations on a or b. First consider the range $-\pi/2<\gamma\leqslant-\pi/4$. Add an infinitesimal $\varepsilon/2$, $0<\varepsilon/2\leqslant\pi/4$ to the boundary value $\gamma=-\pi/2$, so that $\gamma=-\pi/2+\varepsilon/2$. Then

$$N_a = \tfrac{1}{2}[1-\cos\varepsilon/(1-S\sin\varepsilon)]. \quad (15)$$

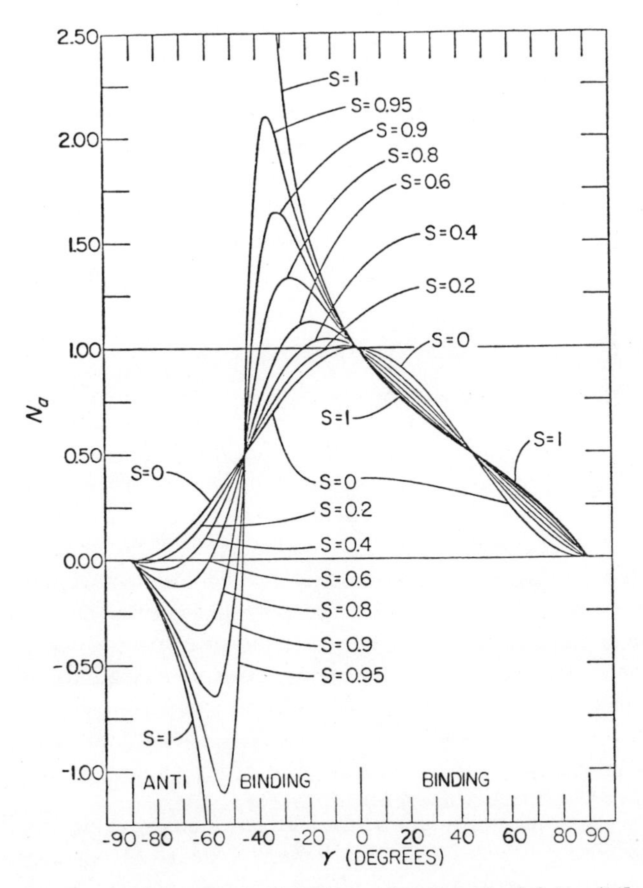

FIG. 4 Gross population at He in HeH^{2+} as a function of γ [cf. Eq. (14)]. [Prepared by K. Ruedenberg and included here with his kind permission.]

Expand $\cos \varepsilon$ and $\sin \varepsilon$ in series and neglect higher-order terms:

$$N_a \approx -\tfrac{1}{2}S\varepsilon/(1-S\varepsilon). \tag{16}$$

Since ε is very small and positive and $0 \leqq S \leqq 1$, the computed N_a is negative. Thus for $-\pi/2 < \gamma \ll -\pi/4$, the computed $N_a < 0$. For $-\pi/2 \ll \gamma < -\pi/4$, the computed $N_a < \frac{1}{2}$ but can be negative, depending on the value of S. In general, in the range $-\pi/2 < \gamma < -\pi/4$, N_a can be negative depending on S. For $\gamma = -\pi/4$, $N_a = \frac{1}{2}$. For $-\pi/4 < \gamma < 0$, the computed N_a is positive and *may* greatly exceed 1, depending on the value of S; if N_a as computed exceeds 1, N_b would be negative. Note that in the ranges $-\pi/2 < \gamma < 0$ just discussed, where Eq. (14) gives anomalous values for N_a, the MO is *anti-bonding* ($b = \sin \gamma$ is negative while $a = \cos \gamma$ is positive). Large S then gives especially great anomalies. Figure 4 (drawn by Professor K. Ruedenberg) shows qualitatively the results just discussed.

For bonding MOs ($0 < \gamma < \pi/2$), there are only slight anomalies, and Eq. (12) can be accepted as substantially correct if a minimal basis, without hybridization in the MOs, is involved. Hybridization or the use of larger basis sets can introduce serious errors even for bonding MOs. It is possible that all these difficulties can be reduced by using MAOs. [See Eq. (3) and the discussion in Section IV.C.] The same considerations as for HeH^{2+} apply to heteropolar diatomic MOs in general.

F. THE NATURE OF COVALENT BINDING

The predominant characteristic physical reasons for covalent bonding are well illustrated by $H_2{}^+$, and will be reviewed here by a qualitative discussion. Strangely enough, these characteristics are commonly misunderstood. At the other extreme, the major features of ionic binding are generally understood.

Hellmann, first proposed that interatomic bonding is due primarily to a lowering of the *kinetic* energy when atoms form a chemical bond [12]. The validity of this idea, and a necessary concomitant, will now be discussed.

From the approximate separation of the Schrödinger equation due to Born and Oppenheimer we obtain the potential curves (or surfaces) with which we commonly work. In the diatomic case, let the curve be called $U(R)$. For stable states, $U(R)$ has a minimum at an equilibrium distance R_e. Now in general

$$U(R) = [T_{el}(R)+V(R)] - [T(\infty)+V(\infty)] = T_{el}(R) + V(R) - E(\infty), \tag{17}$$

where V is the total potential energy, but T_{el} is only the electronic part of the kinetic energy T; E is the total energy. The $U(R)$ curve provides an effective potential for the motions of the nuclei, whose kinetic energy is excluded from $U(R)$.

Equally in quantum and classical mechanics, the following "virial theorem" holds:

$$\overline{T} = -\tfrac{1}{2}\overline{\sum_i x_i F_i}, \qquad \text{where} \quad F_i = -\partial V/\partial x_i; \tag{18}$$

that is, the mean kinetic energy is $-\frac{1}{2}$ times the virial of the forces acting in and on the system. Equation (18) provides relations between T, V, and E (where $E = T+V$) whose form depends on the nature of the forces acting. In an isolated atom or molecule (inverse-square electrostatic forces), it is easily shown (Coulombic virial theorem) that

$$\overline{T} = -\tfrac{1}{2}V = -E. \tag{19}$$

The Coulombic virial theorem cannot be applied to $U(R)$ without taking into account that nuclear kinetic energy T_{nu} has been excluded. $U(R)$ represents the energy only if the nuclei are held fixed, which involves the intervention of external forces. Slater has shown [12] that these lead to the relations

$$\begin{aligned} T_{\text{el}}(R) &= -(E_\infty + U) - R\,dU/dR, \\ V(R) &= 2(E_\infty + U) + R\,dU/dR. \end{aligned} \tag{20}$$

For the *changes* in energy which occur when a molecule is formed without nuclear kinetic energy,

$$\begin{aligned} \Delta T_{\text{el}}(R) &= -U - R\,dU/dR, \\ \Delta V(R) &= 2U + R\,dU/dR. \end{aligned} \tag{21}$$

Application of Eqs. (21) to various situations leads to some interesting results. For stable states, $U(R)$ can be approximated, for R not too small, by

$$U(R) \approx -aR^{-n} + bR^{-m}, \qquad m > n. \tag{22}$$

Then

$$\begin{aligned} \Delta T_{\text{el}} &\approx -(n-1)\,aR^{-n} + (m-1)\,bR^{-m}, \\ \Delta V &\approx (n-2)\,aR^{-n} - (m-2)\,bR^{-m}. \end{aligned} \tag{23}$$

At large R, the first term on the right predominates, hence T decreases and V increases as R decreases. At smaller R, the second term predominates and these relations are reversed. In the typical case $m = 9$, $n = 5$, T reaches a minimum at about $1.2R_e$ and then increases, while V reaches a maximum at slightly larger R, then decreases. [At small R, V again increases, but this behavior is not correctly represented using Eq. (22).] Figure 5 shows the behavior of ΔT_{el} and ΔV as actually computed for three approximations to the wave function of $H_2{}^+$ (see below for details). These show the behavior just stated.

For a repulsive $U(R)$ curve, for example in the interaction of two He

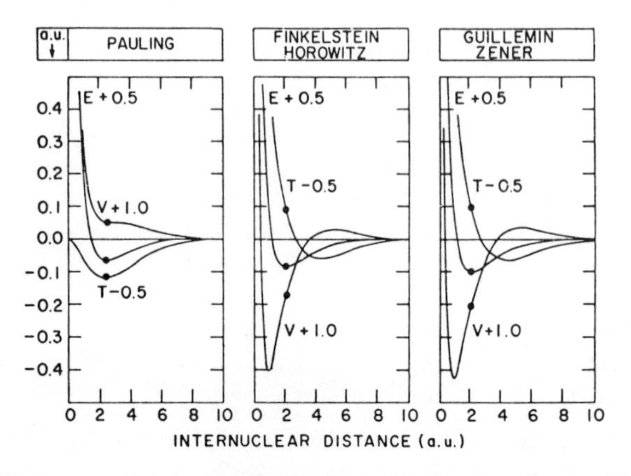

FIG. 5 Kinetic, potential, and total energies of binding for the ground state of $H_2{}^+$, as functions of internuclear distance, for the PL, FH, and GZ wave functions. [From K. Ruedenberg, *in* "Localization and Delocalization in Quantum Chemistry" (O. Chalvet *et al.* eds.), Vol. I, pp. 223–245, Reidel, Dordrecht, 1975.]

atoms, Eqs. (22) and (23) are still valid, but the region where ΔT_{el} is negative and ΔV positive is confined to large R values mostly outside the van der Waals minimum, with ΔT_{el} and ΔV small. At smaller R, ΔT_{el} rises strongly and ΔV falls.

For an ion-pair curve,

$$U(R) \approx -e^2/R, \tag{24}$$

$T_{\mathrm{el}} = 0$, $V = -e^2/R$. In actual ion-pair molecules, however, $U(R)$ is an atom-pair curve for larger R values, until this curve crosses the ion-pair curve. Hence as two atoms (e.g., Na + Cl) approach to form an ionic molecule, T_{el} at first falls and V rises just as for a stable covalent state.

Now following Ruedenberg [13], who with his collaborators has discussed the problem in great detail [14], we approach the structure of the ground state of $H_2{}^+$ in terms of three successive approximations of the LCAO form,

$$(1s_a + 1s_b)/(2+2S)^{1/2}. \tag{25}$$

In the simplest form, first used by Pauling (P), 1s is an AO of the free hydrogen atom: in atomic units,

$$1s = \pi^{-1/2} \exp(-r). \tag{26}$$

In the improved approximation of Finkelstein and Horowitz (FH),

$$1s = (\zeta^3/\pi)^{1/2} \exp(-\zeta r), \tag{27}$$

where ζ is a function of R, chosen so as to minimize the energy at each value of R. Here 1s is an MAO. Still better is the approximation of Guillemin and Zener (GZ),

$$\begin{aligned} 1s_a &= C \exp -[\zeta_1 r_a + \zeta_2 r_b], \\ 1s_b &= C \exp -[\zeta_1 r_b + \zeta_2 r_a], \end{aligned} \tag{28}$$

where the 1s MAOs are polarized forms. Still more accurate would be Eq. (1a).

We shall be interested in the behavior of the potential energy and the electronic kinetic energy as the two nuclei of H_2^+ approach each other along the $U(R)$ curve coming in from large distances toward R_e. Referring to Fig. 5, taken from Ruedenberg's article [13], we see that for any one of the approximate forms of the 1s MAO given by Eqs. (26)–(28), T_{el} falls, and V rises, as the nuclei of H_2^+ approach from large distances. (In Fig. 5, ΔT_{el} and ΔV are shown.) The FH and GZ curves in Fig. 5 have been drawn to correspond to ζ values that minimize $U = T_{el} + V$ (for FH, $\zeta = 1.2387$ at R_e). In the figure, R_e $(= 2.0$ a.u.) is marked with a dot. We ask: Why does V rise, and why does T fall?

For V, the answer can be seen from Eqs. (8)–(10) and Fig. 3. The LCAO expression Eq. (25) involves a transfer of charge from the regions of the two nuclei into the overlap region, increasingly so as R decreases. Most of the overlap region is between the nuclei. The charge density ρ is increased in the region between the nuclei, and decreased elsewhere as a result of the wavelike overlap between $1s_a$ and $1s_b$. Hence, the change in ρ, relative to the sum of the densities $\frac{1}{2}(1s_a{}^2 + 1s_b{}^2)$ for classical overlap, is called the interference density by Ruedenberg. One might think that this transfer of charge into the region between the nuclei would cause a lowering of V because of the transfer from each nucleus to a region closer to the attraction of the other, but a careful qualitative consideration and definitively a calculation shows that the reverse is the case: V is raised [14].

For the kinetic energy, we note that classically, $T = p^2/2m$, where the momentum p is mv. In quantum mechanics, p_z becomes $(h/2\pi i)\, \partial/\partial z$ (let z be along the line joining the two nuclei) and similarly for p_x and p_y; T becomes $-(h^2/8\pi^2 m) V^2$, or in atomic units $-\frac{1}{2}\nabla^2$. Calculation now shows [14] that as a result of the change in charge density distribution, T_{el} is decreased. How can this be understood? The de Broglie wave-mechanical relation $p_z = h/\lambda_z$ holds for the z component of momentum. Now, although we cannot describe the λ of a single H atom, nor of H_2^+, in terms of a single wavelength, it is clear that the effective or mean λ_z is greatly increased in H_2^+ (cf. Fig. 3c) as compared with a single atom. Hence p_z should be decreased, and $p_z{}^2/2m$ and $p^2/2m = T$ also decreased (p_x and p_y are much less affected in the formation of H_2^+). Thus the result of the quantitative calculation on T_{el} is qualitatively understandable.

The preceding discussion is applicable using any one of Eqs. (26)–(28) for the wave function. But as R decreases, what about Eq. (19) ,which must hold (since then $T_{el} = T$) when R_e is reached? Inspection of Fig. 5 shows that it is not fulfilled if Eq. (26) is used for 1s, whereas for Eq. (27) or (28), *provided* the ζ's are properly chosen, it is fulfilled. The major essential point is illustrated by the FH approximation with Eq. (27); use of Eq. (28) or (1a) involves a lesser though of course desirable improvement.

With Eq. (26), V is too high and T is too low. How can this be remedied? The answer is, by increasing ζ as in Eq. (26). This shifts charge closer to the nuclei, lowering V but also increasing T because the average $p^2/2m$ is increased. When, at R_e, ζ is correctly adjusted, as determined by minimizing E, Eq. (19) is satisfied. At R_e, $E = V + T$ is lower than at $R = \infty$, which with $V = -2T$ requires that $V = 2E$ is lower than at ∞ and $T = -E$ is higher than at ∞.

Let us try to understand this in detail. V is lower than at ∞ because the increased ζ has concentrated ρ near the nuclei to an extent *more than sufficient* to overbalance the opposing effect of the transfer of charge from regions around the nuclei to the region between the nuclei. T is higher than at $R = \infty$ as the net result of two partially compensating effects:

(1) T_{el}, in Particular T_z, is decreased in the region between the nuclei because λ_z is increased,

(2) in the regions around the nuclei where the increased ζ increases ρ, T_{el} is increased.

Effect (2) overbalances effect (1), as it must to satisfy the virial theorem Eq. (19) for the total system at R_e, but nevertheless effect (1) contributes to the total. In fact, effect (1) is the key factor which makes it possible for the molecule to be stable; but *only in conjunction with* the freedom to concentrate charge around the nuclei corresponding to the increased ζ. One may ask, which is the cart and which is the horse? According to our discussion, T_{el} is the horse that starts the ζ cart moving. Yet in the end, it is the decrease in V which on balance accounts for the decrease in E.

The relations just discussed for H_2^+ are typical of covalent bonding in general because, usually, the one-electron effects dominate over the two-electron effects with respect to bond formation; for example, in H_2 the inter-electronic repulsion presents only a minor complication which results in the binding energy of the two-electron bond in H_2 being somewhat less than twice the binding energy of H_2^+ [14]. Also, the Pauli antisymmetry principle associated with electron exchange affects the total bonding only in that it limits the number of electrons which can engage in bonding through the same orbital. As in H_2^+, bonding MOs, such as formulated in Eq. (3), show increased ζ for the MAOs relative to the corresponding AOs. Correspondingly, orbital binding energies $-\varepsilon$ and corresponding empirical ionization energies are larger for bonding MOs than for the related AOs. The essential reasons in terms of T and V behavior are the same as in H_2^+.

The foregoing discussion is applicable to covalent bonding in general, not just to homopolar molecules. In heteropolar molecules, the bonding is often largely covalent, and even in molecules usually classed as ionic, there is a little covalent character. The ion-pair case has already been discussed in connection with Eq. (24). It is of interest that for the mildly heteropolar molecule NaLi, also for its positive ion $NaLi^+$, T and V curves similar to those of Fig. 5 have been obtained (See Section VI.F and Ref. 46a in Chapter VI.)

REFERENCES

1. K. B. Jefferts, *Phys. Rev. Lett.* **20**, 39 (1968); **23**, 1476 (1969).
2. See D. R. Bates, K. Ledsham, and A. L. Stewart, *Phil. Trans Roy. Soc. London Ser. A* **246**, 215 (1953).
3. K. Ruedenberg, R. C. Raffenetti, and R. D. Bardo, "Energy Structure and Reactivity" (D. W. Smith, ed.,) p. 164, Wiley, New York, 1973; R. C. Raffenetti, *J. Chem. Phys.* **59**, 5936 (1973); R. D. Bardo and K. Ruedenberg, *ibid.* **59**, 5956, 5966 (1973); R. C. Raffenetti and K. Ruedenberg, *ibid.* **59**, 5978 (1973).
4. (a) R. S. Mulliken, *J. Chem. Phys.* **36**, 3428 (1962); **43**, S39 (1965).
   (b) E. F. Hayes and R. G. Parr, *ibid.* **46**, 3577 (1967).
5. F. Weinhold, *J. Chem. Phys.* **54**, 1874 (1971).
6. R. S. Mulliken, *J. Chem. Phys.* **7**, 14 (1939).
7. D. R. Bates, R. T. S. Darling, S. C. Harre, and A. L. Stewart, *Proc. Phys. Soc. London Sect. A* **66**, 1124 (1953).
8. D. R. Bates, U. Öpik, and G. Poots, *Proc. Phys. Soc. London. Sect. A* **66**, 1113 (1953).
9. D. R. Bates and T. R. Carson, *Proc. Roy. Soc. London Sect. A* **234**, 207 (1156).
10. L. Y. Wilson and G. A. Gallup, *J. Chem. Phys.* **45**, 586 (1966).
11. R. S. Mulliken, *J. Chem. Phys.* **23**, 1833, 1841, 2338, 2343 (1955).
12. H. Hellmann, *Z. Phys.* **85**, 180 (1933); "Quantenchemie," Deuticke, 1937; R. E. Peierls, "Quantum Theory of Solids," p.101, Oxford Univ. (Clarendon) Press, London and New York, (1955); J. C. Slater, *J. Chem. Phys.* **1**, 687 (1933); "Quantum Theory of Molecules and Solids," Vol. 1, Electronic Structure of Molecules, Sections 2-4 and 2-5, McGraw-Hill, New York, 1963. See also R. F. W. Bader and H. J. T. Preston, *Int. J. Quantum Chem.* **3**, 327 (1969).
13. K. Ruedenberg, *in* "Localization and Delocalization in Quantum Chemistry," (O. Chalvet *et al.*, eds.), Vol. 1, pp. 223–245, Reidel, Dordrecht, 1975.
14. K. Ruedenberg, *Rev. Mod. Phys.* **34**, 326–376 (1962); general discussion, especially H_2; M. J. Feinberg, K. Ruedenberg, and E. L. Mehler, *Adv. Quantum Chem.* **5**, 27–98, (1970); on H_2^+ explicitly. For further discussion using a rather different approach see W. A. Goddard, III, and C. W. Wilson, Jr., *Theor. Chim. Acta* **26**, 195, 211 (1972).

CHAPTER III

TWO- TO FOUR-ELECTRON SYSTEMS

A. EXACT CALCULATIONS ON H_2

H_2 is the simplest prototype of polyelectron homopolar diatomic molecules. Here,

$$\mathscr{H} = -\tfrac{1}{2}\nabla_1^2 - \tfrac{1}{2}\nabla_2^2 - \frac{1}{r_{a1}} - \frac{1}{r_{b1}} - \frac{1}{r_{a2}} - \frac{1}{r_{b2}} + \frac{1}{r_{12}}, \tag{1}$$

differing from $\mathscr{H}$ of H_2^+ because there are two electrons, which bring in the new type of complication involved in the interelectronic repulsion term $1/r_{12}$. Solutions of the Schrödinger equation have been obtained for several of the lowest energy states in a form first used by James and Coolidge [1], an expansion in elliptical coordinates together with r_{12}, as

$$\Psi = \sum_{mnjkp} c_{mnjkp} \exp(-\delta[\lambda_1+\lambda_2])(\lambda_1{}^m\lambda_2{}^n\mu_1{}^j\mu_2{}^k\rho^p + \lambda_1{}^n\lambda_2{}^m\mu_1{}^k\mu_2{}^j\rho^p), \tag{2}$$

where λ and μ are as defined for H_2^+ in Chapter II, the subscripts refer to the two electrons, and $\rho = 2r_{12}/R$. The summation is extended over zero and positive values of the indices, subject to the restriction that $j+k$ must be even as required by the nuclear symmetry, and taking as many terms as necessary to give an acceptable approximation to the energy.

With a 100-term function, plus some small corrections, Kolos and Wolniewicz [2] have found, for the ground state and for the ionization potential, energies which are within about 1 cm^{-1} of the experimental values [3]. With calculations at a succession of R values, they have also obtained a theoretical potential curve for the ground state. Their results represent the greatest triumph of *ab initio* calculations. However, since it is not feasible to treat most molecules in the same way, we shall not give further details here, but nevertheless from time to time we shall describe some of the results obtained.

B. N, T, V, Z STATES OF H_2

Figure 1 shows potential curves for several electronic states of H_2. In this section we consider three important states, N, the normal or ground state, T and V, the next lowest states (T refers to triplet), and also a related state called Z. These are valence-shell or, in the case of T, V, and Z at small R, semi-Rydberg or near-Rydberg states. The respective electron configurations near R_e of state N, and the overall species labels, are

$$\begin{aligned} &\text{N:}\quad 1\sigma_g{}^2,\ {}^1\Sigma_g{}^+; &&\text{T:}\quad 1\sigma_g\,1\sigma_u,\ {}^3\Sigma_u{}^+; \\ &\text{V:}\quad 1\sigma_g\,1\sigma_u,\ {}^1\Sigma_u{}^+; &&\text{Z:}\quad 1\sigma_u{}^2,\ {}^1\Sigma_g{}^+. \end{aligned} \tag{3}$$

Higher excited states (Rydberg states) will be dealt with in Section E. The MOs $1\sigma_g$ and $1\sigma_u$ are qualitatively similar to like-named MOs of $H_2{}^+$.

Although the James–Coolidge method yields *accurate* wave functions, from here on we shall use only SCF descriptions, or modifications of these, to take account of electron correlation (i.e., the detailed effects of the r_{12} terms in $\mathscr{H}$). In the James–Coolidge method the latter was dealt with by introducing functions of r_{12} directly into the wave functions. It should be noted that the modifications of the SCF wave function of state N for electron correlation, while minor at R_e, become of major importance as $R \to \infty$. For states T, V, and Z, however, electron correlation in the SCF wave function is less important; in fact, for state T it vanishes as $R \to \infty$. [On the other hand, with the Heitler–London wave function (see below), correlation vanishes as $R \to \infty$ for both states N and T.]

In the SCF approximation the wave functions of states N, T, V, Z are

$$\begin{aligned} \Psi_N^{SCF} &= 1\sigma_g(1)\,1\sigma_g(2)\,S_{12}, \\ \Psi_T^{SCF} &= (1\sigma_g \times 1\sigma_u)\,T_{12}, \\ \Psi_V^{SCF} &= (1\sigma_g \cdot 1\sigma_u)\,S_{12}, \\ \Psi_Z^{SCF} &= 1\sigma_u(1)\,1\sigma_u(2)\,S_{12}. \end{aligned} \tag{4}$$

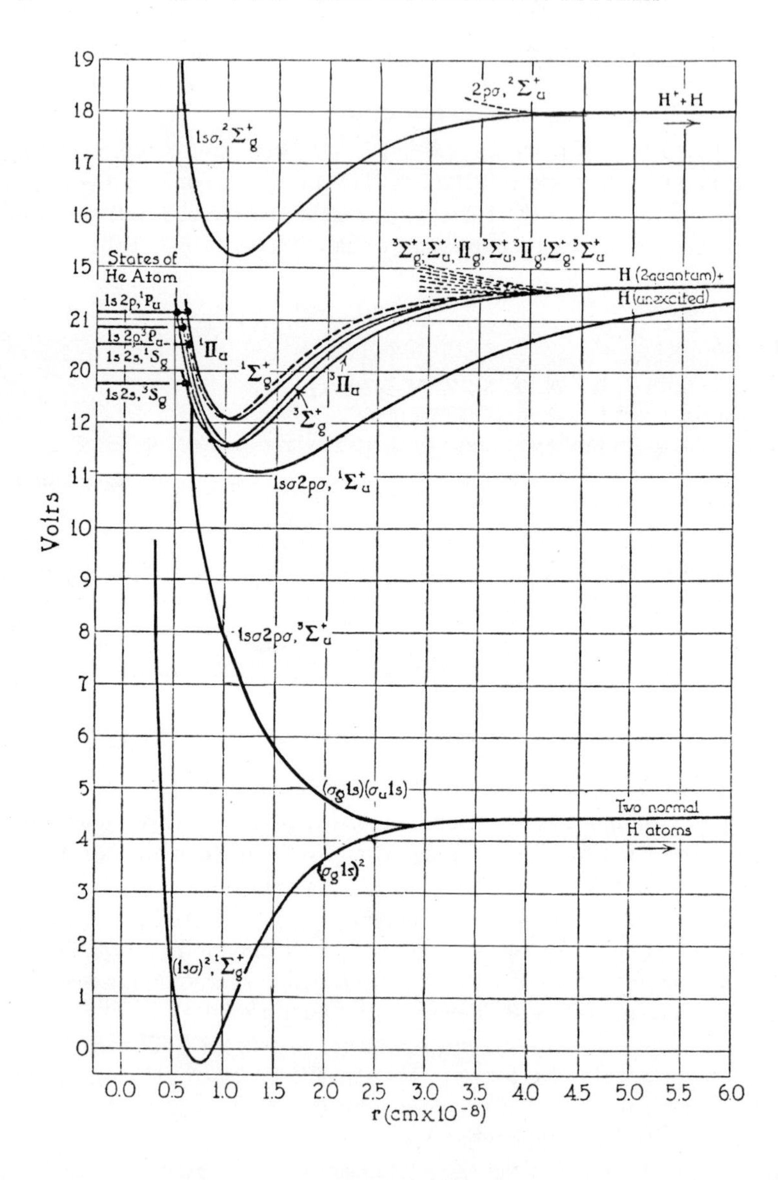

FIG. 1 Lowest H_2 potential curves. [From R. S. Mulliken *Rev. Mod. Phys.* **4**, 46 (1932), except that the $1s\sigma 2p\sigma$, ${}^3\Sigma_u{}^+$ curve has been redrawn.]

In Eqs. (4), introducing a type of notation which will often be used subsequently,

$$\begin{aligned} 1\sigma_g \cdot 1\sigma_u &= [1\sigma_g(1)\,1\sigma_u(2) + 1\sigma_g(2)\,1\sigma_u(1)]/2^{1/2}, \\ 1\sigma_g \times 1\sigma_u &= [1\sigma_g(1)\,1\sigma_u(2) - 1\sigma_g(2)\,1\sigma_u(1)]/2^{1/2}. \end{aligned} \tag{5}$$

Further, S_{12} and T_{12} are respectively singlet spin functions and triplet spin functions:

$$\begin{aligned} S_{12} &= [\alpha(1)\,\beta(2) - \alpha(2)\,\beta(1)]/2^{1/2}, \\ T_{12} &= \alpha(1)\,\alpha(2) \quad \text{or} \quad [\alpha(1)\,\beta(2) + \alpha(2)\,\beta(1)]/2^{1/2} \quad \text{or} \quad \beta(1)\,\beta(2). \end{aligned} \tag{6}$$

The MOs $1\sigma_g$ and $1\sigma_u$ in Eq. (4) depend on R and also differ for the N, T, V, and Z states. Using a *minimal basis set*, $1\sigma_g$ and $1\sigma_u$ are $\sigma_g 1s^\zeta$ and $\sigma_u 1s^\zeta$; that is, $(1s_a{}^\zeta \pm 1s_b{}^\zeta)/2^{1/2}(1 \pm S)^{1/2}$, but their optimized ζ values vary widely, as illustrated in Table 1. In the table, $R = 1.4$ a.u. is R_e of state N, while

TABLE 1

Best Minimal Basis ζ Values for 1s in $1\sigma_g$ and $1\sigma_u$ MOs of H_2

	$R = 1.4$ a.u.	$R = 2$ a.u.
$1\sigma_g$ in N	1.190	1.072
$1\sigma_g$ in T	1.323	1.175
$1\sigma_g$ in V	1.388	1.226
$1\sigma_u$ in T	0.567	0.756
$1\sigma_u$ in V	0.272	0.310

$R = 2$ a.u. is R_e of $H_2{}^+$. We note that with decreasing R, $1\sigma_g$ gets smaller (larger ζ) and $1\sigma_u$ gets larger (smaller ζ), as is required in their approach toward united-atom sizes. State N approaches $1s^2$, ^1S, states T and V approach $1s2p\sigma$, $^3P(\Sigma^+)$ and $^1P(\Sigma^+)$ of the united atom He.

More exactly, using LCSTF expansions, a rather accurate solution of the SCF–MO problem at $R = 1.4$ a.u. yields

$$\begin{aligned} 1\sigma_g(\mathrm{N}) &\approx 0.482\sigma_g 1s^{(1.16)} + 0.036\sigma_g 1s^{(1.84)} \\ &\quad + 0.022\sigma_g 2s^{(1.04)} - 0.032\sigma_g 2p^{(1.67)} + 0.004\sigma_g 3d^{(2.51)}, \\ 1\sigma_u(\mathrm{T}) &\approx 1.86\sigma_u 1s^{(1.11)} + 1.477\sigma_u 1s^{(0.52)} \\ &\quad + 0.016\sigma_u 2s^{(1.80)} - 0.159\sigma_u 2p^{(0.90)}, \\ 1\sigma_u(\mathrm{V}) &\approx -0.517\sigma_u 1s^{(0.75)} + 4.20\sigma_u 1s^{(0.36)} \\ &\quad - 0.054\sigma_u 2s^{(1.80)} - 0.035\sigma_u 2p^{(0.90)}. \end{aligned} \tag{7}$$

The numbers in parentheses are ζ values of 1s, 2s, 2pσ, 3dσ, STFs (not AOs), and $\sigma_g 1s^\zeta$ and so on are LCSTF forms expressed as in Eq. (II.1a).

The computed energies of states N, T, and V using these MOs and including $1/R$ are -1.1336, -0.7775, and 0.6985 a.u. A nearly perfect description of $1\sigma_g$(N) using more STFs than in Eq. (7) is tabulated by Cade and Wahl in a recent publication [4].

As R increases, Ψ_N^{SCF} becomes an increasingly worse approximation to the exact Ψ_N (see below), while Ψ^{SCF} for the T, V, and Z states remains a fairly good approximation. As R increases, the ζ's of Table 1 change, approaching 1 for the T state, which dissociates to two ground-state H atoms as $R \to \infty$, and about 0.7 for the V and Z states [5]. Because of avoided curve intersections, however, the V and Z potential curves, which otherwise would go to H^+ plus H^-, are depressed at very large R values and dissociate to give one normal and one $n = 2$ excited atom [5].

To better understand the N, T, V, and Z states, we need to use Heitler–London wave functions, suitable to describe hydrogen atoms (or ions) approaching from $R = \infty$. These are

$$
\begin{aligned}
\Psi_N^{AO} &\approx (1s_a \cdot 1s_b)\, S_{12}/(1-S^2)^{1/2}, \\
\Psi_T^{AO} &\approx (1s_a \times 1s_b)\, T_{12}/(1+S^2)^{1/2}, \\
\Psi_V^{AO} &\approx [1s_a(1)\, 1s_a(2) - 1s_b(1)\, 1s_b(2)]\, S_{12}/2^{1/2}(1-S^2)^{1/2}, \\
\Psi_Z^{AO} &\approx [1s_a(1)\, 1s_a(2) + 1s_b(1)\, 1s_b(2)]\, S_{12}/2^{1/2}(1-S^2)^{1/2}.
\end{aligned}
\tag{8}
$$

The meanings of the symbols $1s_a \cdot 1s_b$ and $1s_a \times 1s_b$ are like those in Eqs. (4) but $1s_a$ and $1s_b$ here replace $1\sigma_g$ and $1\sigma_u$ there. In Ψ_N^{AO} and Ψ_T^{AO}, 1s refers to H atom orbitals ($\zeta = 1$) and Ψ_V^{AO} and Ψ_Z^{AO} to H^- orbitals ($\zeta \approx 0.7$); the S values differ accordingly in the two cases. [As already remarked, however, Ψ_V and Ψ_Z on dissociation actually go to one normal and one excited atom, yet at moderately large R values, they are well represented by Eqs. (8).]

At large internuclear distances, second-order London dispersion forces produce (as for all other atoms or molecules) a small attractive potential before the Heitler–London forces become appreciable [6].[1] This is true for both the N and T states. On the effects of second-order forces in state N, see also Section V.G.

For excited states dissociating to give one excited atom, *first-order* dispersion forces make rather appreciable, sometimes positive, sometimes negative, contributions to the interatomic potential function [8].

It should be mentioned here that *accurate* calculations by the James–Coolidge method of the energies of the N, T, and V states for a large range

[1] Even the Heitler–London forces make minute qualitatively similar attractive contributions for the T and V states at very large distances [7].

of R values are available [2, 9]. Other properties of state N have also been computed [9b]. Regarding the Z state, see Section E. For the N and T states, Kolos and Wolniewicz [9c] have included the van der Waals region. Here the T state has a minimum of depth 4.49 cm^{-1} at 2.83 a.u.

C. APPLICATIONS OF SCF CALCULATIONS

For state N at R near R_e, the orbital energy ε for $1\sigma_g$ (see Chapter I regarding ε) is a fairly good approximation to the ionization energy I; the *accurate* SCF ε is 16.18 eV, while the experimental I is 15.42 eV. An analogous statement applies to each of the ε's for any SCF wave function of any molecule (Koopmans' theorem; see Chapter I), so long as the SCF wave function is a moderately good approximation.

SCF wave functions are usually fairly good approximations for medium R values, extending up to R considerably greater than R_e. Hence when the electronic SCF energy, plus the nuclear repulsion $1/R$, is computed and plotted as a function of R, a rather accurate potential energy curve $U(R)$ on which molecular vibration can take place is obtained. The minimum of this curve occurs at an R value close to the experimental equilibrium distance R_e for the molecule, while the shape of the curve can be matched in terms of the usual spectroscopic constants ω_e, $x_e\omega_e$, α_e, etc. The agreement with experiment here for H_2, as well as for similarly computed SCF-based $U(R)$ curves for other diatomic molecules, is good but not perfect. Thus for state N of the SCF computed H_2, R_e, ω_e, $\omega_e x_e$, α_e are 0.7276 Å, 4581.7 cm^{-1}, 108.17 cm^{-1}, and 2.719 cm^{-1}, while the experimental values are 0.7417 Å, 4395.2 cm^{-1}, 118.00 cm^{-1}, and 2.993 cm^{-1} [10].

That the computed R_e is a little too small and the computed ω_e a little large for state N can be understood from the fact that $U(R)$ for Ψ^{SCF} for this state rises more rapidly with R because of electron correlation than for the exact Ψ (see Section D). A similar comment is applicable to the ground states of other stable closed-shell molecules. In general, however, for other than ground states, the nature of the discrepancy between SCF $U(R)$ curves and experiment varies from case to case. However, the discrepancies are in general small for the major spectroscopic constants R_e and ω_e, so that SCF computations are very often useful as a guide in identifying states found from experimental spectra.

D. ELECTRON CORRELATION

Electron correlation is required if one wishes to obtain an exact Ψ for any polyelectronic molecule: Ψ^{SCF} may be supplemented by the inclusion of a series of terms belonging to electron configurations different than those for Ψ^{SCF} (see Section I.D). This is configuration mixing (CM). (Note, however, that if the James–Coolidge method can be used, the accurate total energy can be computed directly.) The magnitude of the correlation energy for state N of H_2 as a function of R is of interest. It is almost the same at R_e as for the united atom He: 0.0420 a.u. at $R = 0$, 0.0408 a.u. at R_e. Then it increases, at first slowly, then more rapidly, clearly showing the need of increasing CM with increasing R [11]. This near constancy followed by an increase shows why the SCF-based $U(R)$ curve is nearly but not quite correct in shape near R_e (see Section C).

For state N of H_2 at R_e, the following representation gives about 90% of the correlation energy when natural orbitals (NOs) are used for the MOs [12]:

$$\Psi_N \approx 0.9909\Psi(1\sigma_g{}^2) - 0.1108\Psi(1\sigma_u{}^2) - 0.0551\Psi(2\sigma_g{}^2) - 0.0659\Psi(1\pi_u{}^2) \tag{9}$$

minus smaller terms. That Ψ_N for H_2, provided NOs are used, can be expressed as in Eq. (9) without mixed terms such as $\Psi(1\sigma_g 2\sigma_g)$ was shown in the discussion of NOs in Section I.F, Example (5). This situation is characteristic only for two-electron molecules.

In Eq. (9), the terms involving $1\sigma_u$, $2\sigma_g$, and $1\pi_u$, respectively, take care (to a large extent) of left–right correlation, making the electrons tend to stay near different nuclei; in–out correlation, where the electrons tend to repel each other in such a way that one electron is close to the axis and the other farther out radially; and angular correlation, where the two electrons tend to keep to opposite sides of an axial plane. These effects are illustrated graphically in an instructive discussion by Bader and Chandra [13].

In Eq. (9), $\Psi(1\sigma_g{}^2)$ is approximately Ψ_N^{SCF} of Eq. (4) with $1\sigma_g$ approximately as in Eq. (7). All MOs in Eq. (9) can be expressed in series form as in Eq. (7) plus added terms, with suitably adjusted coefficients and ζ values such that $2\sigma_g$ is roughly $\sigma_g 2s$ and $1\pi_u$ is roughly $\pi_u 2p$ but with ζ values much nearer those in $\sigma_g 1s$ than in excited orbitals of the H atom. This last specification makes all MOs in Eq. (9) roughly the same size, which is a necessity in efficient configuration mixing. The terms in Eq. (9) must all be of ${}^1\Sigma_g{}^+$ symmetry as required for state N. This need is obviously fulfilled by the $\Psi(n\sigma_g{}^2)$ terms and by the $\Psi(1\sigma_u{}^2)$ term. The $\Psi(1\pi_u{}^2)$ term must also be ${}^1\Sigma_g{}^+$, which can be accomplished by constructing it as a suitable linear combination,

$$\Psi(1\pi_u{}^2) = [(1x_u)^2 + (1y_u)^2]/2^{1/2}, \tag{10}$$

where $1x_u \approx \pi_{xu} 2p_x$ and $1y_u \approx \pi_{yu} 2p_y$.

As R increases, the coefficients in Eq. (9) change, and the MOs become modified. As $R \to \infty$, *all the terms vanish* except $\Psi(1\sigma_g{}^2)$ and $\Psi(1\sigma_u{}^2)$. These now have equal coefficients $2^{-1/2}$. Further, $1\sigma_g$ and $1\sigma_u$ acquire exactly the simple forms $(1s_a \pm 1s_b)/2^{1/2}$, with $\zeta = 1$ for both MOs. Then on expanding Ψ_N of Eq. (9) in terms of $1s_a$ and $1s_b$, it is found to be identical with Ψ_N^{AO} of Eq. (8), which is correct for Ψ_N on dissociation of the molecule. It is interesting to note that for large R, Ψ_N corresponds to a strong mixing of the N and Z wave functions of Eq. (8) (50–50 as $R \to \infty$). An indication of this mixing may be found in the fact that the $U(R)$ curve of state N shows a small kink approximately at the R value where the Z state has its minimum.

For diatomic molecules in general, the use of only a few CM terms suffices to permit correct dissociation to atoms. In H_2 only the one added term $(1\sigma_u)^2$ is needed, and dissociation then yields exactly two atoms in their ground state. In general, the method of "optimized valence configurations" (OVC) by Das and Wahl [14] includes at least the minimum number of configurations to allow correct dissociation to SCF ground-state atoms. Now in its most general form, generally called the *multiconfiguration-SCF* (MCSCF) method, a generalized SCF procedure is used [14, 15] which simultaneously optimizes CM coefficients and the forms of the STFs in the LCSTF expansions for all individual MOs used in all CM terms (see Section I.E). The MCSCF procedure in general includes more than enough CM terms to give dissociation to SCF atomic functions, and can be further supplemented to include many more CM terms. In the original OVC method, inner-shell correlation is specifically excluded. In some but not all MCSCF calculations, inner-shell correlation has been included, as is desirable when possible.

In the early days of the recent period of extensive computations, a variety of approximate expressions were developed for obtaining as much as possible of the correlation energy with relatively simple formulas. A summary of such expressions for H_2 is given by McLean *et al.* [16]. For the most part these are now of historical interest only.

The simplest procedure which has been used in gaining a portion of the correlation energy is to replace the two electrons occupying identical orbitals in an electron configuration by electrons in two somewhat different orbitals of the same species; for example,

$$\begin{aligned} &\text{replace} \quad 1s^2 \text{ of He} \quad \text{by} \quad 1s \cdot 1s', \\ &\text{replace} \quad 1\sigma_g{}^2 \text{ of } H_2 \quad \text{by} \quad 1\sigma_g \cdot 1\sigma_g{}', \end{aligned} \tag{11}$$

where 1s and 1s′ differ in ζ, and $1\sigma_g$ and $1\sigma_g{}'$ differ in ζ values and/or in other ways [cf. Eqs. (3) regarding the meaning of 1s·1s′, etc.]. Various writers have used this "different-orbitals-for-different-electrons" (DODE) approach in discussions of atoms and molecules. Now, however, a systematic MCSCF and/or CM approach is generally used instead.

For the H_2 molecule, Coulson and Fischer [17] introduced a description in which $1\sigma_g$ and $1\sigma_g'$ of (11) are replaced by the (here unnormalized) forms ($\chi = 1s$)

$$1\sigma = \chi_a + \lambda\chi_b, \qquad 1\sigma' = \chi_b + \lambda\chi_a. \tag{12}$$

In this formulation, $\lambda \to 0$ as $R \to \infty$, correctly giving the Heitler–London form on dissociation. Near R_e, λ reaches a maximum value (but only about 0.1 when $\zeta = 1$ is used for the χ's; the optimum ζ is about 1.2); the presence of λ here is equivalent to using $c_g\Psi(1\sigma_g^2)+c_u\Psi(1\sigma_u^2)$ without the other terms in Eq. (9). As $R \to 0$, χ_a and χ_b become identical, and $\lambda = 0$; but ζ should be increased.

Goddard, in an extensive series of papers reviewed in Ref. 18 has developed these approaches systematically under the name of the *generalized valence bond* (GVB) method. This method has the advantage, like the OVC and MCSCF methods, of giving correct dissociation behavior, and also functions as a sort of MAO extension [cf. Eq. (II.3)] of the valence-bond method.

E. RYDBERG STATES

Rydberg states [19] are electronic states in which one electron is excited to an MO large in size compared with a usually singly charged "core." The Rydberg states of atoms and molecules fall into series such that the term value [i.e., the energy (in cm^{-1}) to remove the excited electron] is expressible in the form

$$T = \mathrm{Ry}\, Z_c^2/n^{*2} = \mathrm{Ry}\, Z_c^2/(n-\delta)^2, \tag{13}$$

where Z_c is the charge on the core ($Z_c = 1$ for a neutral system) and "Ry" is the Rydberg constant (109,737 cm^{-1} for the H atom). The *quantum defects* δ are usually positive quantities which depend only slightly on the total quantum number n, but often differ strongly for different species of Rydberg MO. For small molecules, the value of n can be assigned as that for the united atom, but for larger molecules, this assignment makes little sense; no agreement has been reached about what do do then, but there is a tendency to use the n of the principal AO component of the LCAO expression for the MO.

In each Rydberg series of an atom or molecule, δ approaches a limiting value δ^∞ as n increases. The occurrence of a substantial positive δ is attributable mainly to penetration of the Rydberg MO into the core, but where this is small, a small positive δ due to polarization becomes important. Exchange effects also contribute; for example, in cases where an electron configuration gives rise to a singlet and a triplet state, the exchange effect makes a negative contribution to δ in the singlet and a positive contribution in the triplet. In atoms, this can in rare cases (e.g., in the 1s3d configuration of He) lead to a small negative total δ for the singlet state. In molecules, the application of Eq. (13) is more complicated than in atoms, for three reasons.

First, there is a *core-splitting* contribution to δ which for some MOs is negative. This effect is strongly manifested in the MOs of H_2^+, which at any given R can be shown [19] all to fall into Rydberg series. These series show δ's, all zero at $R = 0$, which at first increase rapidly in magnitude with R, but in some cases decrease again at large R. Promoted MOs (e.g., $n\text{p}\sigma$) at first show positive δ's. As examples, δ for ns is -0.17, -0.36, and about -0.6 at 1.0, 2.0, and 4.0 a.u. respectively; for $n\text{p}\sigma$, it is $+0.89$, about $+0.2$, and about $+0.1$ at 1.0, 2.0, and 4.0 a.u. These effects are a result of the perturbation of the united-atom field when the united-atom nucleus is split.

In the H_2 molecule, the same effects occur, modified, however, by shielding of the nuclei in the H_2^+ core by the presence of one $1\sigma_g$ electron [20]. The result is a negative contribution to δ in the case of unpromoted MOs. In addition, there is a positive contribution from penetration into the $1\sigma_g$ of the core. The sum total of these effects, plus exchange, often leads to a negative δ for singlet states. Table 2 lists δ values and other data for several states of H_2. A similar table for He_2 shows only positive δ values because of the lesser importance of core-splitting and the greater importance there of the penetration contributions. In molecules with more electrons, the core-splitting effect probably becomes very small.

In Table 2 and hereafter, for H_2 and He_2 we shall designate Rydberg MOs by their united-atom symbols (e.g., $3\text{d}\pi$) rather than by their general (e.g., $1\pi_g$) MO symbols.

A *second* reason why the application of Eq. (13) is more complicated for molecules than for atoms is that in molecules the T value varies with R and, also, at large R values it usually becomes rather meaningless because the SCF wave function without correlation fails to be a good approximation.

A *third* reason is that molecular vibration and rotation introduce complications. Ordinarily one will be most interested in the *vertical* T, taken at R equal to R_e of the positive ion. Other possibilities include T_{00}, corresponding to zero vibration in the ionic and Rydberg states, or T_e, in which the zero-point vibrations have been eliminated from T_{00}. One complication is the fact that, as n increases, electronic–rotational interaction tends toward a Hund's case (d) situation in which two Rydberg series (e.g., in diatomic molecules if the n^* values of a $\text{p}\sigma$ and $\text{p}\pi$ series are close to each other) combine to case (d) series. At very large n values, there is now a single case (d) series for each rotational state of the H_2^+ core, modified by various perturbations [21].

The distinction between valence-shell excited states and Rydberg states is not always clear-cut. In fact, as $R \rightarrow 0$, most excited states sooner or later

TABLE 2

Data on Rydberg States of H_2[a]

Rydberg MO		Triplets				Singlets			
		$n=2$	$n=3$	$n=4$	$n=5$	$n=2$	$n=3$	$n=4$	$n=5$
ns	δ	0.066	0.055			-0.083	-0.091		
	$\Delta G_{1/2}$ (cm^{-1})	2524	2269			2330	2924		
	R_e (Å)	0.989	1.045			1.012			
npσ		(0.67)[a]	0.513	0.473		0.210	0.196	0.187	
		Unstable	2063			1318	1852	2059	
			1.107			1.293	1.134	1.104	
npπ		0.076	0.064	0.062	0.061	-0.081	-0.080	-0.078	-0.079
		2339	2240	2210	2196	2306	2266	2204	2191
		1.038	1.050	1.067	1.057	1.031	1.047	1.061	.1045
ndσ			0.062	0.054		0.052	0.066		
			2088	2149		2232			
ndπ			0.034	0.023		0.022			
			2115	2154		2102	2143		
			1.070			1.069			
ndδ			0.011			-0.035	-0.027		
			2215			2215			
			1.054			1.054			

[a] The δ values correspond to removal of the Rydberg electron from $v = 0$ of the Rydberg state to $v = 0$ of H_2^+ (v is the vibrational quantum number), except that the value for 2pσ, $^3\Sigma_u^+$ is a vertical one for R equal to R_e of H_2^+; $\Delta G_{1/2} = E(v=1) - E(v=0)$ is roughly equal to ω_e.

become Rydberg states [19b]. For the T and V states of H_2 in particular, the transition toward Rydberg character is already well along toward completion even at R_e of the ground state. Such states can be called semi-Rydberg or near-Rydberg states [19]. At large R values, as already noted, the concept of Rydberg states also runs into difficulties, although these are less if the core has a closed-shell structure (as, e.g., in NO). But for the most part, when we discuss Rydberg states, our interest is in cases with R relatively small.

Some Rydberg MOs penetrate strongly into the core; these have *precursors* in the core. Thus in H_2, $1\sigma_g$ is a core MO which is the precursor of the $ns\sigma_g$ Rydberg series, whose first Rydberg member is the $2s(2\sigma_g)$ MO. Here one may speak of $1\sigma_g$ as a *core precursor*, in contrast to an *excited* precursor; in the $ns\sigma_g$ series, all Rydberg MOs with lower n values than the one considered are *excited precursors.* In larger molecules than H_2, several Rydberg series have core precursors; for example, $1\pi_u$ in N_2 ($2p\pi$ in the united atom) is a core precursor of the $np\pi$ Rydberg series.

In molecular Rydberg states, the Rydberg MO is H-atom like in its nonpenetrating parts, except for an inward shift of the radial nodes, increasing with δ [19, 20]. In designating molecular Rydberg series, besides n (or n^*) and other symbols such as σ, π, δ, etc., one must also distinguish series which differ in azimuthal (l) quantum number of the united atom. Thus in H_2 and other diatomic molecules, one has for example both $s\sigma$ and $d\sigma$ Rydberg series. Since $s\sigma$ and $d\sigma$ both belong to the same rigorous symmetry type σ (or σ_g in homopolar diatomics), however, they are not pure $s\sigma$ or $d\sigma$ but somewhat mixed except as $R \to 0$. Nevertheless, one has both predominantly $s\sigma$ and predominantly $d\sigma$ series, with different major precursors; for example, in N_2, $1\sigma_g$ and $2\sigma_g$ are major precursors of the $s\sigma_g$ series, and $3\sigma_g$ of the $d\sigma_g$ series; or perhaps $2\sigma_g$ and $3\sigma_g$ should both be regarded as in part precursors of both the $ns\sigma_g$ and $nd\sigma_g$ series.[2] In H_2, the $s\sigma_g$ series has $1\sigma_g$ as precursor, but the $d\sigma_g$ series has no core precursor; however, there must in both be some slight mixing of s and $d\sigma$.[3]

Calculations of the James–Coolidge type on several of the Rydberg states of H_2, with resulting $U(R)$ curves, have been made by Davidson and coworkers [24] ($1\sigma_g 2s$, $1\sigma_g 3s$, $1\sigma_g 3d\sigma$, $^3\Sigma_g^+$, $1\sigma_g 2p\pi$, $^1\Pi_u$, $1\sigma_g 3s$, $^1\Sigma_g^+$, $1\sigma_g 3d\sigma$, $^1\Sigma_g^+$, $1\sigma_g 3d\delta$, $^3\Delta_g$ and $^1\Delta_g$), and somewhat more accurately by Kolos and Wolniewicz for the $1\sigma_g 2s$, $^3\Sigma_g^+$ and $1\sigma_g 2p\pi$, $^1\Pi_u$ states, and by Kolos and Rychlewski on the $1\sigma_g 2p\pi$ and $1\sigma_g 3p\pi$, $^1\Pi_u$ states [25]. Although these accurate calculations include electron correlation, they do

[2] The core MOs $2\sigma_g$ and $3\sigma_g$ in N_2 can both be expressed as mixtures of semi-united-atom 2s and $3d\sigma$ AOs, but $2\sigma_g$ is more nearly 2s and $3\sigma_g$ more nearly $3d\sigma$ (cf. Mulliken [22]).

[3] See Wakefield and Davidson [23] for some relevant calculations on the $1\sigma_g 3s$, $^1\Sigma_g^+$, and $1\sigma_g 3d\sigma$, $^1\Sigma_g^+$ states of H_2.

not go very far beyond SCF in accuracy. This is because near R_e and at smaller values the wave functions of Rydberg states can be expressed to a good approximation as simple antisymmetrized products of a core (here H_2^+) function times a Rydberg MO occupied by the outer electron [24]. Rothenberg and Davidson [24] give an instructive analysis in terms of NOs.

The potential curve of the $1\sigma_g 2s$, ${}^1\Sigma_g^+$ Rydberg state interacts strongly with that of the Z state (see Section A) to give a single curve with two minima. The form of this $U(R)$ curve has been computed using a James–Coolidge type of wave function by Davidson, and more accurately by Kolos and Wolniewicz [26].

As $R \to 0$, MOs approach their united-atom forms, and for Rydberg MOs in H_2 the approach is already close at R_e. This approach is illustrated by some approximate SCF calculations of Mulliken [27] on the $1\sigma_g 3d\pi$, ${}^3\Pi_g$ Rydberg state of H_2 at $R = 2$ a.u. (close to R_e). Let the $1\pi_g(3d\pi)$ MO be approximated as

$$C_1(2p\pi_a - 2p\pi_b) + C_2\, 3d\pi_c + C_3(3d\pi_a + 3d\pi_b), \tag{14}$$

where the subscript c refers to an STF centered at the midpoint of the molecule. With optimized ζ's (0.257 for $2p\pi$ and 0.340 for $3d\pi$), several energy-minimizing calculations were made setting various coefficients in Eq. (14) equal to zero, with the results shown in Table 3.[4] The $3d\pi$ population (see Section II.E.1) of the $1\pi_g$ MO was also computed for each case. For $1\sigma_g$, a $\sum_i c_i \chi_i$ STF expression

$$1s^{(1.10)} - 0.12\, 2s^{(1.00)} + 0.07\, 2p\sigma^{(1.56)} + 0.008\, 3d\sigma^{(2.5)}$$

TABLE 3

H_2 Molecule $1\sigma_g 1\pi_g$ ${}^3\Pi_g$ State

C_1	C_2	C_3	$-\varepsilon(1\pi_g)$	$-E$	Pop. $3d\pi$
0.842	−0.265	−0.284	0.05628	0.658239	0.82
1.006	0	−0.401	0.05628	0.658236	0.78
0.526	−0.886	0	0.05627	0.658220	0.89
0	0.894	0.054		0.658198	1.00
0	1.000	0	0.05624	0.658197	1.00
0	0	0.512	0.05617	0.658126	1.00
4.608	0	0	0.05487	0.656823	0

[4] Actually the results in Table 3 correspond to less well-optimized ζ's, namely 0.245 for $2p\pi$ and 0.341 for $3d\pi$, but serve to illustrate the points made here. With the better ζ's, E in the second line of Table 2 becomes −0.658427.

was used for each nucleus, with slight energy-minimizing variations in the coefficients for the different cases.

Obviously $1\pi_g$ is better approximated by $3d\pi_a + 3d\pi_b$, or especially by the united-atom form $3d\pi_c$, than by $2p\pi_a - 2p\pi_b$, but the last of these alone is seen to do surprisingly well; however, this is true only when the ζ for $2p\pi$ has been decreased very strongly from its H-atom value of 0.5. In any event, it is seen that the inclusion of some $2p\pi_a - 2p\pi_b$ is helpful even though $3d\pi_c$ gives the best result for the use of a single term. Also of interest is the fairly close approach of $-\varepsilon(1\pi_g)$, here 0.05617 to the value 0.05556 which it has in the H atom. As $R \to 0$, this value is reached at about $R = 0.2$ a.u.

Although correlation between core and Rydberg MOs is a minor effect for Rydberg states for R near or less than R_e (see above), it becomes increasingly important for *even-electron molecules* including H_2, as $R \to \infty$. In other words, the neat separation between core and Rydberg electron breaks down. (In odd-electron molecules, for example NO, the separation between the now usually closed-shell core and the Rydberg electron remains valid all the way to $R = \infty$.) So long as the core and Rydberg electron remain nearly independent, the $U(R)$ curve of the Rydberg state, and so the spectroscopic constants of that state, are nearly the same as for the core, namely H_2^+ in the case of H_2 (cf. Table 2). This behavior continues to perhaps about $1.5R_e$, after which correlation has become important.

If the Rydberg state of H_2 can be well described at R_e by an SCF function of configuration $1\sigma_g x$, Ψ must undergo two kinds of changes as $R \to \infty$:

(1) the x MO goes into an LCAO form $y+y$ or $y-y$,
(2) the wave function goes toward $1\sigma_g(y \pm y) + \lambda\, 1\sigma_u(y \mp y)$, with $\lambda \to \pm 1$ as $R \to \infty$, and with $1\sigma_g \to 1s + 1s$ and $1\sigma_u \to 1s - 1s$.

For $\lambda = 1$ we attain the covalent Heitler–London-like state $H + H^*(y)$, while for $\lambda = -1$ we have the ion-pair state $H^+ + H^-$ $(1s \cdot y)$, or since excited H^- is probably unstable, just $H^* + H(1s) + e$. If x is an unpromoted MO, for example ns, $np\pi$, or $nd\delta$, $y = x$, and the upper signs are appropriate. If x is a promoted MO, $y \neq x$, and the lower signs are appropriate; for example, if x is $np\sigma$, then $y = (n-1)s$; if x is $nd\sigma$, y is $(n-1)p\sigma$; if x is $nd\pi$, y is $(n-1)p\pi$; if x is $nf\sigma$, then $y = (n-2)p\sigma$ [28].

In cases where x is an unpromoted MO, $U(R)$ curves of fairly normal shape result as $R \to \infty$. These curves show stable minima considerably lower than their asymptotes; in other words, they have appreciable positive dissociation energies. At large R values, however, such curves often show *small* maxima. Thus in the case of $1\sigma_g 2p\pi$, ${}^1\Pi_u$, there is a small maximum [24, 25, 29] which is due to first-order dispersion forces [30].

When x is a promoted MO, relatively *large* maxima occur because the

$U(R)$ curve first rises following approximately the H_2^+ curve, as R increases, then falls because dissociation occurs to give an H-atom state with relatively low energy [31]. The occurrence of these humps has been confirmed by theoretical calculations on the $1\sigma_g 3d\pi$, $^3\Pi_g$ state by Wright and Davidson [24], and on this and the related $^1\Pi_g$ state by Browne [32]. Similar curves with humps occur in the He_2 molecule, as shown in Fig. 3.in Section I.

F. SPECTROSCOPIC TRANSITION PROBABILITIES

The probability of a spectroscopic transition between two electronic states m and n depends on the dipole moment μ_{mn}^q of the transition (cf. Section II.C). Of especial interest in H_2 are the absorption and emission transitions between the ground state and the excited states V($1\sigma_g 1\sigma_u$, $^1\Sigma_u^+$: Lyman bands) and $1\sigma_g 2p\pi$, $^1\Pi_u$ (Werner bands).

The electronic transition moments can be computed by Eq. (II.4). To obtain exact values, exact Ψ's must be used, but it is of interest first to see what can be obtained using Ψ^{SCF} expressions. Then

$$\mu_{NE}^q = \int \Psi_N^{SCF}(q_1+q_2)\Psi_E^{SCF}\,dv,$$

where N refers to the ground and E to the excited state. Putting $\Psi_N^{SCF} = (1\sigma_g{}^N)^2 S_{12}$ and $\Psi_E^{SCF} = 1\sigma_g{}^E \cdot \phi^E S_{12}$ [cf. Eq. (3)], it is easily shown that

$$\mu_{NE}^q = 2^{1/2}\int 1\sigma_g{}^N 1\sigma_g{}^E\,dv \int 1\sigma_g{}^N q\phi^E\,dv, \tag{15}$$

where ϕ^E is $1\sigma_u$ or $1\pi_u$ for E = V or $^1\Pi_u$ respectively, and $q = z$ for E = V, or x and y, or $x \pm iy$, for E = $^1\Pi_u$.

For the V–N transition, Eq. (15) gives the same approximate results as Eq. (II.5b) for H_2^+ if we use LCMAO expressions for $1\sigma_g$ and $1\sigma_u$, except that now there is an additional factor $2^{1/2}\int 1\sigma_g{}^N 1\sigma_g{}^V\,dv$, or approximately $2^{1/2}$. When we consider the transition probability, proportional to $(\mu_{NV})^2$, this makes a factor of about 2, which is explained by the fact that two electrons are now available for the transition between $1\sigma_g$ and $1\sigma_u$, instead of just 1 as in H_2^+. For a more exact SCF calculation of the V–N Lyman transition moment, expressions such as Eq. (7) can be substituted in Eq. (15), and the integrations performed; note that the result will vary strongly with R. Similar calculations can be made for the Werner transition; there (see below) the result is nearly independent of R.

Using the *exact* wave functions of Kolos and Wolniewicz, Wolniewicz has computed the electronic transition moments of the Lyman and Werner bands with the results shown in Fig. 2, taken from a paper by Wolniewicz [33]. For Fig. 2, it should be kept in mind that μ_{NE} for the Werner bands

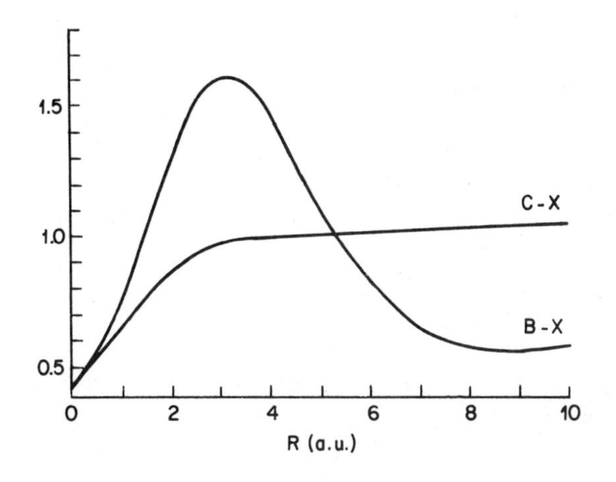

FIG. 2 Electronic transition moments, $\mu_{nm}e$, for the Lyman (B–X) and Werner (C–X) bands. [From L. Wolniewicz, *J. Chem. Phys.* **51**, 5002 (1969).]

applies to either the x or the y component of the transition moment. Hence in using Fig. 2 to calculate transition probabilities (each proportional to the *square* of the transition moment), a factor 2 must be introduced for the intensity of the Werner bands in order to allow for both components. Making use of the electronic transition moments as computed by Wolniewicz, Allison and Dalgarno [34] after bringing vibrational wave functions into the computation have obtained transition probabilities and oscillator strengths for the individual *bands* of the Lyman and Werner systems. For absorption from the zero vibrational level of the ground state of H_2, the sum total oscillator strength is 0.31 for the Lyman bands and 0.36 for the Werner bands. The result for the Lyman bands is not very different from values obtained in early calculations using approximate SCF wave functions [35].

G. POLARIZABILITY OF H_2

Given its wave function, many properties of a molecule can be computed. (See Section C with regard to some of these for the SCF wave function.) Of considerable interest is the polarizability, for which the differing values along ($\alpha_{\parallel}$) and perpendicular ($\alpha_{\perp}$) to the internuclear axis and their average (α) can be computed by applying perturbation theory on consideration of the effects of an external field on the wave function. Using their wave functions, Kolos and Wolniewicz have carried out the necessary calculations [36].

They have tabulated the computed $\alpha_{\parallel}$, $\alpha_{\perp}$, and α as a function of R from

$R = 0.4$ to 4.0 a.u., and have also tabulated α and γ $(= \alpha_{\parallel} - \alpha_{\perp}$, the anisotropy of the polarizability) as a function of the vibrational and rotational quantum numbers. From the latter data, the theoretical value of α for H_2 gas as a function of temperature can be computed. We note also that the *experimental* α can be obtained from the dispersion formula which gives the refractive index as a function of frequency, taken for frequency zero.

From α and γ, a theoretical value of the depolarization ratio for Rayleigh scattering was calculated and compared with experiment. From the nondiagonal matrix elements of α and γ for vibrational quantum numbers 0 and 1, the depolarization ratio for Raman scattering was also calculated and compared with experiment. An absolute intensity for Raman scattering was calculated from the anisotropy of polarizability and compared with experiment.

H. H_2^-, H_2+H, H_3^+, H_3^-, AND H_3^{2+}

When H and H^- approach, one can compute the beginning of a potential curve corresponding to the ground state $1\sigma_g^2 1\sigma_u$, ${}^2\Sigma_u^+$ of H_2^-, isoelectronic with the stable ground state of He_2^+ (see Section I). But the computed H_2^- potential curve before long reaches an energy equal to that of H_2 in its ground state plus a free electron, and the electron will then fly off [37]. The same is true of the $1\sigma_g 1\sigma_u^2$, ${}^2\Sigma_g^+$ first excited (repulsive) state. Among other excited states, however, many short-lived "resonances" occur [38].

Although no really stable H_3 radical exists (however, there should be a dispersion force attraction between H_2 and an H atom), the activation barrier for the reaction

$$H_2 + H \rightarrow H + H_2$$

is of great interest to chemists [39]. It has been shown that the minimum energy approach of H to H_2 is by way of a linear configuration of the three nuclei. The most accurate calculation available is that by Liu using SCF MOs plus CM [39b]. The calculation yields a linear symmetric saddle point with nearest H–H distance $R = 1.757$ a.u. and an energy above that of free H_2+H of 10.28 kcal/mole, with an estimate that the accurate value is not less than 9.8 kcal/mole. Liu believes that the methods used are applicable to larger systems.

Qualitative considerations are already sufficient to show that the equilibrium geometry is that of an equilateral triangle (D_{3h}) for the ground states of H_3^+ and H_3^{2+}, with electron configurations $(1a')^2$ and $1a'$ respectively. The best SCF calculation gives an energy for H_3^+ of -1.2993 a.u. [40]; when CM is added, a value of -1.3397 a.u. is obtained [41], and the estimated accurate energy is -1.342 ± 0.001 a.u. The computed equilibrium H–H separation is 1.66 a.u. Schaad and Hicks [42] give CM calculations on a number of excited states. These are mostly unstable, of varying geometries.

In the chemical problem of the formation of H_3^+, two routes are possible: from $H_2^+ + H$, or, lower in energy when the parts are separated, from $H_2 + H^+$. The energy of the path is minimal for an approach of the odd atom or ion along a line perpendicular to the line joining the nuclei in H_2 or H_2^+. The energy curves for the two processes cross during the approach, however, and the correct surface must be obtained from an avoided intersection. The necessary *ab initio* calculations have been carried out [43] using CM. This example is the first calculation of a situation of this kind.

The molecule H_3^- is unstable except for a small polarization binding at large separations. However, rather good CM calculations on the system have been carried out [44]. The energy is minimized by a linear arrangement of the nuclei.

I. THE He_2 MOLECULE AND ITS IONS

Although two He atoms do not form a stable molecule, He_2^+ is stable, and a large number of stable Rydberg states of He_2 are known, largely from emission spectroscopy. These excited states of He_2 belong to the class of excimer states, a term which is used for states formed by the combination of an unexcited molecule with an excited molecule of the same species.

Several *ab initio* calculations have been made on the ground state of He_2^+, of which the most accurate to date, using LCSTF functions plus extensive correlations, is by Liu [45]. For this state Liu computes a (vibrationless) dissociation energy of 2.454 eV, but estimates the accurate value, after allowing for all correlation energy, to be 2.469 ± 0.006 eV. Since the spectroscopically known excited states [46] of He_2 fall into Rydberg series, accurate ionization energies are known for all or most of these states. From these and the computed energy of He_2^+, the absolute energies of the excited states relative to the energy of two He atoms can be obtained.

Of especial interest for experimental reasons are the two lowest excited states,

$$1\sigma_g^2 1\sigma_u 2s,\ {}^3\Sigma_u^+ \text{ and } {}^1\Sigma_u^+.$$

Experimental evidence, and a series of theoretical calculations[5] on the potential curves of these states, indicate that each of these curves has a small hump ("nonobligatory hump," cause not clear) [47b] at a large R value. Using his results on He_2^+, Liu estimates the height of the hump in the ${}^1\Sigma_u^+$ state as 0.080 ± 0.016 eV. Guberman and Goddard find 0.06 eV [47b]. The most accurate theoretical calculation [48a][6] on the ${}^3\Sigma_u^+$ state gives a hump

[5] See Mukamel and Kaldor [47a]. Guberman and Goddard [47b], report a hump of 0.06 eV in the ${}^1\Sigma_u^+$ state and also one of 0.22 eV in the $1\sigma_g^2 1\sigma_u 2\sigma_u$, ${}^1\Sigma_g^+$ state.

[6] For further details on the He_2 states, see Brown and Matsen [48b].

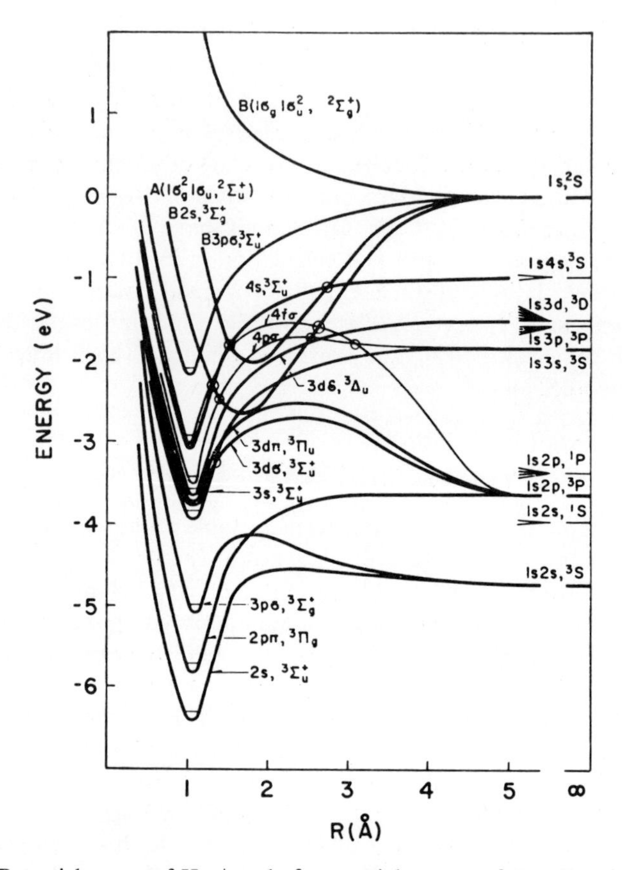

FIG. 3 Potential curves of He_2^+ and of some triplet states of He_2. For the ten lower states shown, which have *A* core, the curve shapes near their minima and their depths relative to the *A* curve of He_2^+ are based on experimental data, except for 4pσ and 4fσ. Otherwise the curves are only qualitatively correct. [From R. S. Mulliken, *Phys. Rev. A* **136**, 962 (1964).]

of 0.11 eV height. Many of the other He_2 states show much larger ("obligatory") humps, whose explanation is analogous to those in some of the Rydberg states of H_2 (see end of Section E). Figure 3 is a qualitative representation of some of the observed and predicted potential curves of He_2, with several examples of humps in the curves. Similar humps must occur for H_2 (see Section E).

In addition to the Rydberg states with a stable $1\sigma_g{}^2 1\sigma_u$, ${}^2\Sigma_u{}^+$ core, addition of a Rydberg electron to He_2^+ in the repulsion state $1\sigma_g 1\sigma_u{}^2$, ${}^2\Sigma_g{}^+$ can be predicted to give a number of stable states [31].

For the ground state of He_2, with a repulsive potential curve except for a slight van der Waals attraction at large R, many computations have been made.[7] When both partners in forming a molecule or complex are closed-shell systems,[8] as here, the use of an SCF MO wave function gives a fairly good approximation as a function of R all the way to $R = \infty$, unlike the case of two partners which form a bond as they approach as in the case of two H atoms forming H_2 (see Section B). In the present case, this can be seen from the fact that the SCF wave function, here $1\sigma_g^2 1\sigma_u^2$, ${}^1\Sigma_g^+$, becomes *identical* as $R \to \infty$ with the antisymmetrized product $1s_a^2 1s_b^2$ of two He atom functions, when use is made of the fact that $1\sigma_g$ and $1\sigma_u$ in the limit have the forms $(1s_a \pm 1s_b)/2^{1/2}$. The identity mentioned can be verified by writing the SCF MO function with the MOs in this form, expanding, and comparing. The CEs (correlation energies) at $R = \infty$ are then also obviously identical. At smaller R values, the $1\sigma_g$ and $1\sigma_u$ take on more complicated LCSTF forms [cf. Eq. (7)], but the changes are continuous.

Thus at smaller R values the *difference* between the energy of He_2 and that of two He atoms is equal to the difference of their respective SCF energies corrected by the *difference* in the CEs of the two systems, which in this case may be assumed to be small because in both cases we are dealing with two closed shells (of MOs in the one case, AOs in the other). The first part of the statement just made is also valid for the approach of two H atoms, but in that case the formation of the $1\sigma_g^2$ closed shell in the molecule requires a substantial new CE correction which was not present for the two atoms ("molecular extra CE").

Following the principle just outlined, Kestner [50a][9] has made an accurate SCF calculation for He_2 at a series of R values down to 1.0 a.u., and by comparing it with an equally accurate SCF calculation on 2 He obtains data for a rather good curve for the repulsion of two He atoms. Kestner found that a basis set consisting only of ζ-optimized 1s, 2s, 2s′, and $2p\sigma$ for each nucleus gives almost the same results as a much more extensive set. Earlier, Phillipson made a calculation for He_2 as a function of R including extensive CM [51a].[10]

One failing of all SCF MO calculations is that they cannot yield the van der Waals dispersion force attractions which exist between any two atoms or molecules at large distances. This failing can be remedied by suitable CM corresponding to the simultaneous excitation of the two atoms involved (especially, both to 1s2p in the two atoms in He_2), as prescribed in the theory

[7] For references through 1973, see Richards *et al.* [49].

[8] Note that the same relation is true for two half-open-shell partners, as in the T state of H_2 (cf. Section B), and nearly so for an open shell plus a half-open shell, as in He_2^+.

[9] Compare also Gilbert and Wahl [50b] for a less complete and slightly less accurate calculation.

[10] Compare also Kunick and Kaldor [51b], who use another method.

of dispersion forces. *Ab initio* calculations along these lines have been made by Bertoncini and Wahl [52a][11] using an MCSCF procedure (cf. Section D), also by Schaefer *et al.* [53a].[12] The results agree fairly well with experimental evidence. The same method should be useful for other diatomic molecules and for long-range intermolecular interactions.

As $R \to 0$, the large-R electron configuration of ground-state He_2 goes over into $1s^2 2p\sigma^2$ of the united atom Be, equivalent to a mixture of one-third ^1S and two-thirds ^1D of the configuration $1s^2 2p^2$. However, the ground-state configuration of this atom is $1s^2 2s^2$, corresponding to $1\sigma_g{}^2 2\sigma_g{}^2$ of He_2. The two computed curves (see Davidson *et al.* [51]) cross at about $R = 0.6$ a.u., and CM is essential to obtain a good wave function which goes properly to the ground-state united-atom function. CM with $1\sigma_g{}^2 1\pi^2$ is also important, while for accuracy lesser amounts of other CM are of course needed. In larger diatomic molecules, it is normal that the ground-state wave function at larger R goes as here to an excited configuration of the united atom. The curves for $1\sigma_g{}^2 2\sigma_g{}^2$ and $1\sigma_g{}^2 1\sigma_u{}^2$ have been extended to $R = 0$ in SCF calculations by Yarkony and Schaefer [54], resulting in correlation diagrams showing the energies of the MOs as functions of R. Both united-atom and separate-atom, and combined, basis sets were used. CM calculations for the region near the avoided crossing are also reported.

The ground state of the two-electron ion He_2^{2+} is of special interest since in the formation of the molecule, after an initial rise because of the mutual repulsion of the two He^+ ions, there is a dip corresponding to the binding which occurs in the H_2 molecule. Details, including results of a minimal-set SCF calculation plus configuration mixing and of a nearly accurate 40-term James–Coolidge-type calculation [55], are given in a paper by Fraga and Ransil [56]. Results of calculations on some excited states of He_2^{2+} are given by Browne [57].

For the one-electron molecule He_2^{3+}, Wilson and Gallup [58] have reported computed curves for the ground state and a number of excited $^2\Sigma$ states. They used elliptical coordinates.

J. HeH, HeH^+; $He + H_2$; $H_2 + H_2$; H_4^+

The HeH radical in its ground state is unstable, but computations disclose that like He_2, it must have many stable excited states with potential curves very similar to those of the positive ion [59]. For HeH^+, the ground state is stable. Many *ab initio* computations are available.[13] The most accurate

[11] For similar calculations on the lowest $^3\Sigma^+$ state, see Das [52b].

[12] There are, however, some difficulties; see the further calculations of Liu and McLean [53b].

[13] A survey of the literature and some computations are given by Yamada *et al.* [60].

computation is one of the James–Coolidge type by Kolos and Peek using an 83-term function [61]. This gives $R_e = 1.4632$ a.u. and $E = -2.97869074$ a.u.; the vibrationless dissociation energy D_e is 16455.64 cm^{-1}. Energies were computed, however, for R values ranging from 0.9 to 4.5 a.u., also for $R = 5$–9 a.u. Dissociation yields $He + H^+$. The first excited state, correlating with $He^+ + H$, and some of the other excited states, probably have small minima, while others are repulsive [62]. Peyerimhoff made an SCF calculation using STF expansions for the MOs; she gives instructive charge-density maps [63]. Green *et al.* have made SCF plus CM calculations on a number of $^1\Sigma^+$ excited states [64].

There are a few papers on the potential surface of the system $H_2 + He$. Gordon and Secrest have made two rather accurate computations [65]. One was of the SCF type, which should give fairly good results since both He and H_2 are closed-shell systems (see the discussion in Section H). In the other, CM was used both for the combined system and for the separate systems. The results for the potential surface were very similar in the two computations. The computations were made for various approach directions ranging from end-on to an approach of the He atom toward the mid-point of the H_2. Electron density plots are given; the H_2 molecule tends to contract when the He atom approaches.

Approximate calculations on the interaction of two H_2 molecules as a function of distance and orientation have been made by various methods. Especially instructive is a small-basis LCGTF plus CM study which makes it possible to calculate both the short-range repulsive components (essentially given by SCF) and the long-range dispersion attractions (essentially given by CM) [66].[14] It is shown that both increase on stretching *either* or (about twice as much) *both* of the H_2 molecules involved. The repulsive forces are minimized by an approach of the two molecules in perpendicular planes [67], although for small separations the energy is lowest for a linear symmetrical arrangement of the four atoms.

Calculations by valence-bond plus CM methods show considerable attraction (up to about 1 eV) between H_2 and H_2^+ for various geometries [68]. However, H_4^+ is barely stable with respect to dissociation into $H_3^+ + H$ (calculated dissociation energy almost 380 cm^{-1}).

LiH, although a four-electron system, will be considered in Chapter IV.

REFERENCES

1. H. M. James and A. S. Coolidge, *J. Chem. Phys.* **1**, 825 (1933).
2. See W. Kolos and L. Wolniewicz, *J. Chem. Phys.* **41**, 3663 (1964); **43**, 2429 (1965); **49**, 404 (1968); **51**, 1417 (1969).

[14] Earlier references are given in this reference; see especially V. Magnasco and G. F. Musso, in part with R. McWeeny.

3. G. Herzberg, *J. Mol. Spectros.* **33**, 147 (1970); G. Herzberg and C. H. Jungen, *ibid.* **41**, 425 (1972).
4. P. E. Cade and A. C. Wahl, *Atomic Data* **13**, 339 (1974).
5. P. E. Phillipson and R. S. Mulliken, *J. Chem. Phys.* **28**, 1248 (1958); **33**, 615 (1960); W. Kolos and L. Wolniewicz, *ibid.* **45**, 509 (1966).
6. J. O. Hirschfelder, C. F. Curtiss, and R. B. Bird, "Molecular Theory of Gases and Liquids." Wiley, New York and Chapman and Hall, London, 1954; 2nd corr. ptg., 1964.
7. W. A. Bingel, H. Preuss, and H. H. Schmidtke, *Z. Naturforsch.* **16a**, 434 (1961); C. Herring and M. Flicker, *Phys. Rev.* **134A**, 362 (1964).
8. R. S. Mulliken, *Phys. Rev.* **120**, 1674 (1960).
9. (a) W. Kolos and L. Wolniewicz, *J. Chem. Phys.* **45**, 509 (1966); **48**, 3672 (1968).
   (b) W. Kolos and L. Wolniewicz, *J. Chem. Phys.* **41**, 3674 (1964).
   (c) W. Kolos and L. Wolniewicz, *Chem. Phys. Lett.* **24**, 457 (1974).
10. A. D. McLean, *J. Chem. Phys.* **40**, 243–244 (1964).
11. W. Kolos and C. C. J. Roothaan, *Rev. Mod. Phys.* **32**, 219 (1960).
12. S. Hagstrum and H. Shull, *Rev. Mod. Phys.* **35**, 624 (1963); E. R. Davidson and L. L. Jones, *J. Chem. Phys.* **37**, 2966 (1962); W. D. Lyons and J. O. Hirschfelder, *ibid.* **46**, 1788 (1967).
13. R. F. W. Bader and A. K. Chandra, *Can. J. Chem.* **46**, 953 (1968), Fig. 11.
14. G. Das and A. C. Wahl, *J. Chem. Phys.* **44**, 87 (1966); **47**, 2934 (1967); **56**, 1769 (1972).
15. E. Clementi, *J. Chem. Phys.* **46**, 3842 (1967); T. L. Gilbert, *Phys. Rev.* **6A**, 580 (1972); J. Hinze, *J. Chem. Phys.* **59**, 6424 (1973).
16. A. D. McLean, A. Weiss, and M. Yoshimine, *Rev. Mod. Phys.* **32**, 211 (1960); R. S. Mulliken, *Proc. Natl. Acad. Sci. USA* **38**, 160 (1952).
17. C. A. Coulson and I. Fischer, *Phil. Mag.* **40**, 386 (1949).
18. W. A. Goddard, III, T. H. Dunning, Jr., W. J. Hunt, and P. J. Hay, *Acc. Chem. Res.* **6**, 368 (1973).
19. R. S. Mulliken, *J. Am. Chem. Soc.* **86**, 3183 (1964); *Acc. Chem. Res.* **9**, 7 (1976).
20. R. S. Mulliken, *J. Am. Chem. Soc.* **91**, 4615 (1969).
21. G. Herzberg and Ch. Jungen, *J. Mol. Spectros.* **41**, 425 (1972); U. Fano, *J. Opt. Soc. Am.* **65**, 979 (1975).
22. R. S. Mulliken, *Int. J. Quantum Chem.* **1**, 103 (1967).
23. C. B. Wakefield and E. R. Davidson, *J. Chem. Phys.* **43**, 834 (1965).
24. W. M. Wright and E. R. Davidson, *J. Chem. Phys.* **43**, 840 (1965); C. B. Wakefield and E. R. Davidson, *ibid.* **43**, 834 (1965); S. Rothenberg and E. R. Davidson, *ibid.* **44**, 730 (1966); **45**, 2560 (1966).
25. W. Kolos and L. Wolniewicz, *Chem. Phys. Lett.* **1**, 19 (1967); *J. Chem. Phys.* **48**, 3672 (1968); W. Kolos and J. Rychlewski, *J. Mol. Spectros.* **62**, 846 (1976).
26. W. Kolos and L. Wolniewicz, *J. Chem. Phys.* **50**, 3228 (1969); F. L. Tobin and J. Hinze, *ibid.* **63**, 1034 (1975).
27. R. S. Mulliken, *Chem. Phys. Lett.* **14**, 141 (1972).
28. R. S. Mulliken, *J. Am. Chem. Soc.* **88**, 1849 (1966).
29. J. C. Browne, *J. Chem. Phys.* **40**, 43 (1964).
30. R. S. Mulliken, *Phys. Rev.* **120**, 1674 (1960).
31. R. S. Mulliken, *Phys. Rev. A***136**, 962 (1964).
32. J. C. Browne, *Phys. Rev. A***138**, 9 (1965).
33. L. Wolniewicz, *J. Chem. Phys.* **51**, 5002 (1960).
34. A. C. Allison and A. Dalgarno, *Atomic Data* **1**, 289–304 (1970).
35. S. Ehrenson and P. E. Phillipson, *J. Chem. Phys.* **34**, 1224 (1961).

36. W. Kolos and L. Wolniewicz, *J. Chem. Phys.* **46**, 1426 (1967).
37. H. S. Taylor and F. E. Harris, *J. Chem. Phys.* **39**, 1012 (1963); P. G. Burke, *J. Phys. B. (Proc. Phys. Soc.)* **1**, 586 (1968).
38. G. J. Schultz, *Rev. Mod. Phys.* **45**, 378 (1973).
39. (a) I. Shavitt, R. M. Stevens, F. L. Minn, and M. Karplus, *J. Chem. Phys.* **48**, 2700 (1968).
    (b) B. Liu, *J. Chem. Phys.* **58**, 1925 (1973).
40. M. E. Schwartz and L. J. Schaad, *J. Chem. Phys.* **47**, 5325 (1967).
41. I. G. Csizmadia, R. E. Karl, J. C. Polanyi, A. C. Roach, and M. A. Robb, *J. Chem. Phys.* **52**, 6205 (1970).
42. L. J. Schaad and W. V. Hicks, *J. Chem. Phys.* **61**, 1934 (1974).
43. C. W. Bauschlicher, Jr., S. V. O'Neil, R. K. Preston, H. F. Schaefer, III, and C. F. Bender, *J. Chem. Phys.* **59**, 1286 (1973).
44. A. Macias, *J. Chem. Phys.* **48**, 3464 (1968); **49**, 2198 (1968).
45. B. Liu, *Phys. Rev. Lett.* **27**, 1251 (1971).
46. M. L. Ginter, *in* "Données spectroscopiques concernant les molecules diatomic" (B. Rosen, ed.), Hermann, Paris, 1970.
47. (a) S. Mukamel and U. Kaldor, *Mol. Phys.* **22**, 1107 (1971).
    (b) S. L. Guberman and W. A. Goddard, III, *Chem. Phys. Lett.* **14**, 460 (1972); *Phys. Rev. A* **12**, 1209 (1975).
48. (a) H. J. Kolker and H. H. Michels, *J. Chem. Phys.* **50**, 1762 (1969).
    (b) J. C. Browne and F. A. Matsen, *Adv. Chem. Phys.* **23**, 161 (1973).
49. W. G. Richards, T. E. H. Walker, and R. K. Hinkley "A Bibliography of *Ab Initio* Molecular Wave Functions." Oxford Univ. (Clarendon) Press, London and New York, 1971, and its supplement for 1970–1973.
50. (a) N. R. Kestner, *J. Chem. Phys.* **48**, 252 (1968).
    (b) T. L. Gilbert and A. C. Wahl, *J. Chem. Phys.* **47**, 3425 (1967).
51. (a) P. E. Phillipson, *Phys. Rev.* **125**, 1981 (1962); G. H. Matsumoto, C. F. Bender, and E. R. Davidson, *J. Chem. Phys.* **46**, 402 (1967).
    (b) D. Kunick and U. Kaldor, *J. Chem. Phys.* **56**, 1741 (1972).
52. (a) P. Bertoncini and A. C. Wahl, *Phys. Rev. Lett.* **25**, 991 (1970); *J. Chem. Phys.* **58**, 1259 (1973).
    (b) G. Das. *Phys. Rev. A* **11**, 732 (1975).
53. (a) H. F. Schaefer, III, D. R. McLaughlin, F. E. Harris, and B. J. Alder, *Phys. Rev. Lett.* **25**, 988 (1970); D. R. McLaughlin and H. F. Schaefer, III, *Chem. Phys. Lett.* **12**, 244 (1971).
    (b) B. Liu and A. D. McLean, *J. Chem. Phys.* **59**, 4557 (1973).
54. D. R. Yarkony and H. F. Schaefer, III, *J. Chem. Phys.* **61**, 4921 (1974); W. C. Ermler, R. S. Mulliken, and A. C. Wahl, *J. Chem. Phys.* **66**, 3031 (1977).
55. W. Kolos and C. C. J. Roothaan, *Rev. Mod. Phys.* **32**, 219 (1960).
56. S. Fraga and B. J. Ransil, *J. Chem. Phys.* **37**, 1112 (1962).
57. J. C. Browne, *J. Chem. Phys.* **42**, 1428 (1965); W. V. Hicks, L. J. Schaad, and K. K. Innes, *ibid.* **65**, 463 (1976).
58. L. Y. Wilson and G. A. Gallup, *J. Chem. Phys.* **45**, 586 (1966).
59. H. H. Michels and F. E. Harris, *J. Chem. Phys.* **39**, 1464 (1963); C. F. Bender and E. R. Davidson, *J. Phys. Chem.* **70**, 2675 (1966).
60. T. Yamada, H. Sato, E. Ishiguro, and T. Takezawa, *J. Phys. Soc. Japan.* **32**, 1595 (1972).
61. W. Kolos and J. M. Peek, *Chem. Phys.* **12**, 381 (1976); W. Kolos, *Int. J. Quantum Chem.* **10**, 217 (1976).
62. H. H. Michels, *J. Chem. Phys.* **44**, 3834 (1966).

63. S. Peyerimhoff, *J. Chem. Phys.* **43**, 998 (1965).
64. T. A. Green, H. H. Michels, J. C. Brown, and M. M. Madsen, *J. Chem. Phys.* **61**, 5186, 5198 (1974).
65. M. D. Gordon and D. Secrest, *J. Chem. Phys.* **52**, 120 (1970); M. D. Gordon, D. Secrest, and C. Llaguno, *ibid.* **55**, 1046 (1971).
66. O. Tapia and G. Bessis *Theor. Chim. Acta.* **25**, 130 (1972).
67. O. Tapia, G. Bessis, and S. B. Bratoz, *Int. J. Quantum. Chem.* **S4**, 289 (1971); J. C. Raich, A. B. Anderson, and W. England, *J. Chem. Phys.* **64**, 5088 (1976).
68. R. D. Poshusta and D. F. Zetik, *J. Chem. Phys.* **58**, 118 (1973).

CHAPTER IV

DIATOMIC HYDRIDES

A. SCF CALCULATIONS

The case of HeH^+ has already been discussed in Section III.J. Very accurate SCF calculations on diatomic hydrides have been made by Cade and Huo [1].[1] A sampling of their results, for the ground states of the molecules LiH, CH, HF, and HCl, is given in Tables 1–4. Their basis sets comprise STFs found suitable for free atoms, somewhat reoptimized after adding ζ-optimized polarization functions. In Tables 1–4, their ζ values and MO coefficients have been rounded off to fewer figures for the sake of perspicacity and with negligible effects on accuracy. Cade and Huo estimate that their computed SCF energies are within 0.001 a.u. of the limiting possible accurate SCF results for the first-row hydrides, but less accurate (but not more than 0.005 a.u. above the accurate results) for the second-row hydrides, because only one dσ function and no fσ functions were used for the latter, by reason of limitations of machine capacity.

In the tables, T and V denote kinetic and potential energies, the latter including nuclear repulsion energy. It should be noted that Tables 1–4 refer to wave functions computed at the experimental R_e values, which are 0.02–

[1] For a list including other calculations through 1973, see Richards *et al.* [2].

TABLE 1

SCF Wave Function for LiH ($1\sigma^2 2\sigma^2$, $^1\Sigma^+$)[a]

STF_{χ_m}	C_{im}: $C_{1\sigma,m}$	$C_{2\sigma,m}$
$1s_{Li}$ (2.521)[b]	0.894	−0.128
$1s'_{Li}$ (4.699)	0.103	−0.004
$2s_{Li}$ (0.797)	−0.003	0.346
$2s'_{Li}$ (1.200)	0.003	−0.032
$3s_{Li}$ (2.750)	0.020	−0.033
$2p\sigma_{Li}$ (0.737)	−0.001	0.176
$2p\sigma'_{Li}$ (1.200)	−0.004	0.046
$2p\sigma''_{Li}$ (2.750)	−0.002	0.006
$3p\sigma_{Li}$ (3.200)	−0.006	−0.004
$3d\sigma_{Li}$ (0.642)	−0.000	0.019
$3d\sigma'_{Li}$ (1.200)	−0.001	0.021
$4f\sigma_{Li}$ (0.925)	−0.000	0.007
$1s_H$ (0.888)	0.007	0.601
$1s_H'$ (1.566)	0.000	0.100
$2s_H$ (2.200)	−0.000	−0.006
$2p\sigma_H$ (1.376)	0.002	0.017

[a] $R = 3.015$ a.u. $= R_e$(exptl);
$E = -7.98731$ a.u., $T = 7.99132$ a.u.,
$V = 15.97863$ a.u., $V/T = 1.9995$;
$\varepsilon_{1\sigma} = -2.4453$ a.u., $\varepsilon_{2\sigma} = -0.3017$ a.u.

[b] ζ Values in parentheses.

0.04 a.u. less than the theoretical SCF values, because of the absence of electron correlation in the latter (see Section III.C). If either an *exact* Ψ is used at the true *experimental* R_e value, or alternatively an SCF Ψ at the *SCF calculated* R_e value, the virial ratio V/T must be exactly 2. At the SCF-computed R_e values (not listed here), V/T is 2.00000 ± 0.00001 in all Cade and Huo's calculations, although the V/T values in the tables are slightly different because of departure of the experimental from the SCF R_e's.

Cade and Huo in their papers give computed results not only for R_e as in Tables 1–4, but for a large number of R values. In each table, the basis set of STFs (index m) is given, followed by the coefficients C_{im} of the several STFs in the LCSTF expressions $\sum_m C_{im}\chi_m$ for the MOs ϕ_i. From these data they compute SCF potential curves and spectroscopic constants which they compare with experiment. As expected (see Section II.C), the agreement is good but not perfect.

In their recent publication in *Atomic Data*, Cade and Huo [1] include accurate SCF calculations not only on the neutral hydrides but also on the unipositive and negative ions of many of them. Other authors also have reported calculations on the NeH^+ [3a] and ArH^+ [3b] ions. On the heavier

TABLE 2

SCF Wave Function for CH($1\sigma^2 2\sigma^2 3\sigma^2 1\pi$, $^2\Pi$)[a]

STF_{χ_m}	C_{im}: $C_{1\sigma,m}$	$C_{2\sigma,m}$	$C_{3\sigma,m}$	STF	C_{im}: $C_{1\pi,m}$
$1s_C$ (5.007)[b]	0.925	−0.217	0.117	$2p\pi_C$ (1.021)	0.423
$1s_C'$ (9.049)	0.128	−0.004	0.001	$2p\pi_C'$ (1.611)	0.457
$2s_C$ (1.298)	−0.002	0.394	−0.345	$2p\pi_C''$ (2.790)	0.168
$2s_C'$ (2.068)	0.003	0.473	−0.263	$2p\pi_C'''$ (6.711)	0.009
$3s_C$ (6.057)	−0.057	−0.025	0.008	$3d\pi_C$ (1.583)	0.033
$2p\sigma_C$ (1.039)	−0.000	0.074	0.372	$4f\pi_C$ (2.076)	−0.005
$2p\sigma_C'$ (1.726)	−0.002	0.131	0.256	$2p\pi_H$ (1.447)	0.032
$2p\sigma_C''$ (2.742)	0.003	0.024	0.125	$3d\pi_H$ (2.726)	0.003
$2p\sigma_C'''$ (6.543)	0.001	0.003	0.007		
$3d\sigma_C$ (1.239)	−0.001	0.036	0.064		
$3d\sigma_C'$ (2.349)	0.000	0.017	0.016		
$4f\sigma_C$ (1.590)	−0.000	0.011	0.010		
$1s_H$ (1.342)	0.005	0.022	0.357		
$1s_H'$ (2.898)	−0.001	0.088	−0.020		
$2s_H$ (2.112)	−0.002	0.175	−0.024		
$2p\sigma_H$ (2.226)	0.000	0.019	0.009		

[a] $R = 2.124 = R_e$(exptl);
$E = -38.27935$, $T = 38.25498$, $V = -76.53433$, $V/T = -2.00064$;
$\varepsilon_{1\sigma} = -11.3162$, $\varepsilon_{2\sigma} = -0.8290$, $\varepsilon_{3\sigma} = -0.4552$, $\varepsilon_{1\pi} = -0.4150$.
[b] ζ Values in parentheses.

metal hydrides, we know of only one SCF calculation, a study of the $^7\Sigma^+$ ground state and a $^7\Pi$ excited state of MnH, by Bagus and Schaefer [4].

BeH at its R_e obviously has the electron configuration and state $1\sigma^2 2\sigma^2 3\sigma$, $^2\Sigma^+$. For a molecule dissociating to a closed-shell atom (Be) plus one electron (in H) as here, one would anticipate (just as for the approach of two closed-shell atoms), that the MO configuration would remain unchanged out to dissociation. In SCF calculations extending over a wide range of R values, however, Mulliken found *two* SCF curves which intersect sharply at 4.26 a.u. (see Fig. 1) [5]. Near this R value, *either* of the two solutions can be obtained from the SCF procedure depending on whether one approaches 4.26 a.u. from smaller or larger R values. The "inner" curve (smaller R) is clearly $1\sigma^2 2\sigma^2 3\sigma$ with 2σ a B–H bonding MO and 3σ an s–p hybrid non-bonding MO. For the "outer" curve, the configuration is best described as $1\sigma^2 2\sigma 3\sigma^2$ with 2σ predominantly a $1s_H$ orbital, becoming pure $1s_H$ as $R \to \infty$, while 3σ is now predominantly a 2s beryllium orbital, becoming pure $2s_{Be}$ as $R \to \infty$; the orbital energy ε approaches the value -0.5 a.u. for 2σ and the numerically smaller SCF value appropriate to $2s_{Be}$ for 3σ, corresponding to dissociation into H 1s, ^2S plus Be $1s^2 2s^2$, ^1S.

TABLE 3

SCF Wave Function for HF$(1\sigma^2 2\sigma^2 3\sigma^2 1\pi^4, {}^1\Sigma^+)$[a]

STF_{χ_m}	C_{im}: $C_{1\sigma,m}$	$C_{2\sigma,m}$	$C_{3\sigma,m}$	STF_{χ_m}	C_{im}: $C_{1\pi,m}$
$1s_F$ (7.944)[b]	0.952	−0.266	0.062	$2p\pi_F$ (1.358)	0.335
$1s_F'$ (14.109)	0.086	0.005	−0.002	$2p\pi_F'$ (2.329)	0.488
$2s_F$ (1.935)	−0.001	0.467	−0.153	$2p\pi_F''$ (4.261)	0.256
$2s_F'$ (3.256)	0.004	0.584	−0.153	$2p\pi_F'''$ (9.297)	0.011
$3s_F$ (9.925)	−0.041	−0.025	0.004	$3d\pi_F$ (2.134)	0.025
$2p\sigma_F$ (1.407)	−0.000	0.040	0.285	$4f\pi_F$ (2.794)	0.007
$2p\sigma_F'$ (2.373)	−0.000	0.059	0.414	$2p\pi_H$ (1.771)	0.026
$2p\sigma_C''$ (4.278)	0.001	0.015	0.204	$3d\pi_H$ (3.320)	0.003
$2p\sigma_C'''$ (8.973)	0.001	0.002	0.011		
$3d\sigma_F$ (1.835)	−0.000	0.019	0.043		
$3d\sigma_F'$ (3.368)	0.000	0.005	0.012		
$4f\sigma_F$ (2.700)	−0.000	0.007	0.012		
$1s_H$ (1.373)	0.001	−0.037	0.186		
$1s_H'$ (2.460)	−0.000	0.076	0.052		
$2s_H$ (2.461)	0.000	0.051	0.041		
$2p\sigma_H$ (2.923)	0.000	0.011	0.011		

[a] $R = 1.7328 = R_e$(exptl);
$E = -100.07030$, $T = 100.02718$, $V = -200.09748$, $V/T = -2.00043$;
$\varepsilon_{1\sigma} = -26.29428$, $\varepsilon_{2\sigma} = -1.60074$, $\varepsilon_{3\sigma} = -0.76810$, $\varepsilon_{1\pi} = -0.65008$.
[b] ζ Values in parentheses.

This example illustrates a situation which for SCF calculations can recur, sometimes in more complicated ways, in other cases. Anomalies of this sort are, of course, immediately resolved when one introduces electron correlation. In the case of BeH, the preceding MO description corresponds to a valence-bond theory description in which there is an avoided crossing of two potential curves, one, dominant at small R, derived from a $1s^2 2s2p$ excited state of Be, the other, dominant at large R, corresponding to a $1s^2 2s^2$ ground-state Be plus $1s_H$.

As has been noted earlier, SCF MO calculations, even though they usually give good approximations at R_e, are usually no longer good at large R values and on dissociation. However, there are some exceptions among ground states (notably, for the combination of any two closed-shell atoms or molecules) and among some excited states. The prototype case is perhaps that of the $1\sigma_g 1\sigma_u$, ${}^3\Sigma_u{}^+$ repulsion state (T state) of H_2 (Section III.A). A rather similar case is the $1\sigma^2 2\sigma^2 3\sigma 4\sigma$, ${}^3\Sigma^+$ state of BH. This too is a repulsion state, except that at small R its potential curve reaches a maximum, then descends to a small minimum with an R_e close to that of BH^+, where the state is clearly a Rydberg state; details will be given in Section E.2. As $R \rightarrow 0$,

TABLE 4

SCF Wave Function for HCl($1\sigma^2 2\sigma^2 3\sigma^2 1\pi^4 4\sigma^2 5\sigma^2 2\pi^4$, X $^1\Sigma^+$)[a]

STF_{χ_m}	C_{im}: $C_{1\sigma,m}$	$C_{2\sigma,m}$	$C_{3\sigma,m}$	$C_{4\sigma,m}$	$C_{5\sigma,m}$
$1s_{Cl}$ (18.673)[b]	0.845	−0.237	0.001	0.067	−0.020
$2s_{Cl}$ (5.794)	0.002	0.929	−0.003	−0.409	0.093
$2s'_{Cl}$ (16.428)	0.184	−0.119	0.000	0.046	−0.013
$3s_{Cl}$ (1.715)	0.000	−0.001	0.000	0.331	−0.229
$3s'_{Cl}$ (2.792)	−0.001	0.008	0.001	0.685	−0.217
$3s''_{Cl}$ (10.116)	−0.000	0.146	−0.001	0.011	0.010
$2p\sigma_{Cl}$ (5.267)	0.000	0.003	0.676	−0.025	−0.161
$2p\sigma'_{Cl}$ (8.325)	−0.000	0.000	0.325	−0.008	−0.067
$2p\sigma''_{Cl}$ (14.021)	0.000	0.000	0.033	−0.001	−0.007
$3p\sigma_{Cl}$ (1.389)	0.000	0.000	−0.001	0.007	0.244
$3p\sigma'_{Cl}$ (2.514)	−0.000	0.001	0.008	0.065	0.507
$3d\sigma_{Cl}$ (2.400)	0.000	0.001	0.001	0.023	0.067
$1s_H$ (1.508)	−0.001	−0.001	0.000	0.214	0.878
$1s_H'$ (2.568)	0.000	0.001	0.000	−0.025	−0.250
$2s_H$ (2.270)	0.000	0.000	−0.000	−0.003	−0.202
$2p\sigma_H$ (1.763)	0.000	−0.000	−0.000	0.030	0.043

STF_{χ_m}	C_{im}: $C_{1\pi,m}$	$C_{2\pi,m}$
$2p\pi_{Cl}$ (5.292)	0.680	−0.195
$2p\pi'_{Cl}$ (8.333)	0.320	−0.080
$2p\pi''_{Cl}$ (13.937)	0.034	−0.009
$3p\pi_{Cl}$ (1.424)	−0.001	0.511
$3p\pi'_{Cl}$ (2.591)	0.008	0.575
$3d\pi_{Cl}$ (2.630)	0.001	0.009
$2p\pi_H$ (1.243)	−0.000	0.046
$3d\pi_H$ (1.811)	−0.000	0.012

[a] $R = 2.4087 = R_e$(exptl);
$E = -460.1103$, $T = 460.0947$, $V = -920.2050$, $V/T = -2.00003$;
$\varepsilon_{1\sigma} = -104.84844$, $\varepsilon_{2\sigma} = -10.57409$, $\varepsilon_{3\sigma} = -8.04195$, $\varepsilon_{4\sigma} = -1.11683$, $\varepsilon_{5\sigma} = -0.62620$,
$\varepsilon_{1\pi} = -8.03936$, $\varepsilon_{2\pi} = -0.47620$.

[b] ζ Values in parentheses.

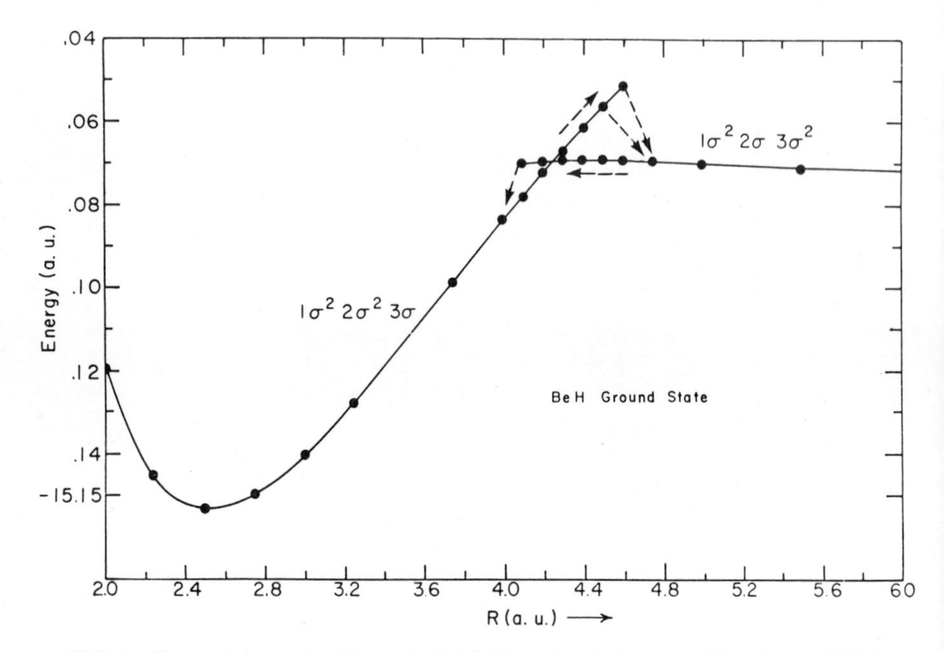

FIG. 1 Computed energies. [From R. S. Mulliken, *Int. J. Quantum Chem.* **5**, 95 (1971).]

the T state of H_2 must also become a Rydberg state, but there is no evidence of a minimum in its potential curve near R_e of $H_2{}^+$. The same is true of the $1\sigma^2 2\sigma^2 3\sigma 4\sigma$, ${}^3\Sigma^+$ state of CH^+ (see Section E.3), even though this is isoelectronic with the ${}^3\Sigma^+$ state of BH. The ground state of BeH already discussed is an interesting anomalous case.

States which become Rydberg states as $R \rightarrow 0$ have been called "Rydbergescent" by Mulliken [6]. Those for which a single SCF configuration remains valid as $R \rightarrow \infty$ are called "MO-dissociating." Evidently such a state may or may not have a Rydberg minimum ($dU/dR = 0$) near R_e of the positive ion. Mulliken [7][2] has examined several MO-dissociating states of CH and NH and has found one state of CH, $1\sigma^2 2\sigma^2 3\sigma^2 4\sigma$, ${}^2\Sigma^+$ where the SCF calculation indicates the presence of a small Rydberg minimum. Of the Rydbergescent states of NH, none showed a Rydberg minimum, but all showed points of inflection near R_e of NH^+; here d^2E/dR^2 is 0, then becomes positive with increasing R, goes over a maximum, and approaches zero as $R \rightarrow \infty$.

Besides calculations on various properties mentioned in the following,

[2] Strictly speaking, the CH state mentioned dissociates into $1s_H$ plus carbon $1s^2 2s^2 2p\sigma^2$, ${}^1\Sigma^+$, which is a mixture of two atomic eigenstates ${}^1D(\frac{2}{3})$ and ${}^1S(\frac{1}{3})$ of the $1s^2 2s^2 2p^2$ configuration.

there have been a number of relatively early calculations on magnetic properties (magnetic susceptibility, rotational moments, spin rotation constants, magnetic shielding at each nucleus), in particular on LiH, BH, and HF [2]. Calculations have also been made on electrical polarizability.

The computation of electronic transition moments and oscillator strengths from SCF wave functions, although these are one-electron properties, in general gives unsatisfactory results. The reason is that *two* electronic wave functions are involved, and each has in general very different correlation terms. Thus in some calculations on hydrides, Henneker and Popkie [8] report for the best-known transition in OH ($^2\Sigma^+ \leftarrow {}^2\Pi$) a calculated oscillator strength which differs by a factor of 2.5 from experiment. When electron correlation is introduced, however, good agreements are obtained (unpublished work of Yoshimine *et al.*).

Scott and Richards [9a] have made approximate SCF calculations on low-lying states of ScH, TiH, and VH. (For more accurate MCSCF and CM calculations on VH, see Section E.6.) Internuclear separations and vibration frequencies are predicted. The ground state of TiH is calculated to be $^4\Phi$.

Julienne *et al.* [9b] have made SCF calculations of the potential curves of the lowest-lying states X $^2\Pi$ and $^2\Sigma^+$ of HF^+, both of which dissociate to H^+ plus F so that SCF calculations are reasonably good approximations out to dissociation. The results are applied to an analysis of the photoelectron spectrum of HF.

B. CHARGE DISTRIBUTIONS

In a series of five papers, Bader, Cade *et al.* have discussed molecular charge distributions and chemical bonding. Three of these papers refer to first-row [10a] and second-row [10b] diatomic hydrides or their ions.[3] Using Cade and Huo's SCF wave functions at R_e values, they compute contour maps of the charge distributions for individual MOs and for the total charge density (in atomic units) in a plane passing through the nuclei. They also give *difference* distributions between total charge densities in molecules and corresponding (a) separate atoms and (b) united atoms. The former difference distributions are very instructive in showing what happens in molecule formation. The molecular maps and the latter difference distributions are also instructive in showing the increasing resemblances of molecules to the united atom in going from LiH to HF. The various contour maps for first-row hydrides are reproduced here in Figs. 2–6. They are worthy of careful study, but since they then speak for themselves, no special discussion will be given here. Since total charge densities at R_e are relatively

[3] For a general review on molecular charge distributions, see Bader [10c].

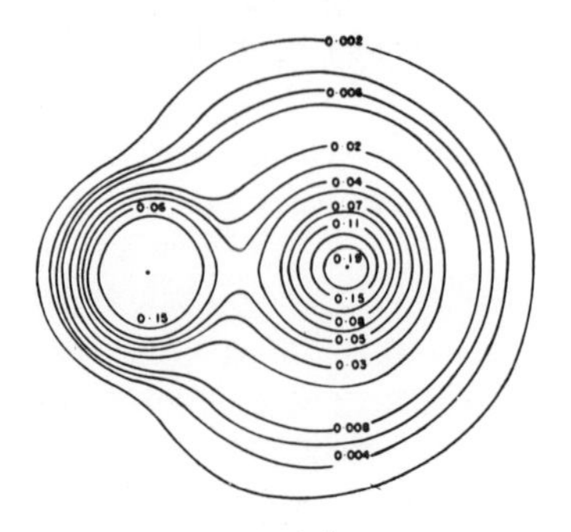
0·002
0·006
0·02
0·04
0·07
0·11
0·06
0·15
0·15
0·05
0·03
0·008
0·004
LIH ¹Σ⁺

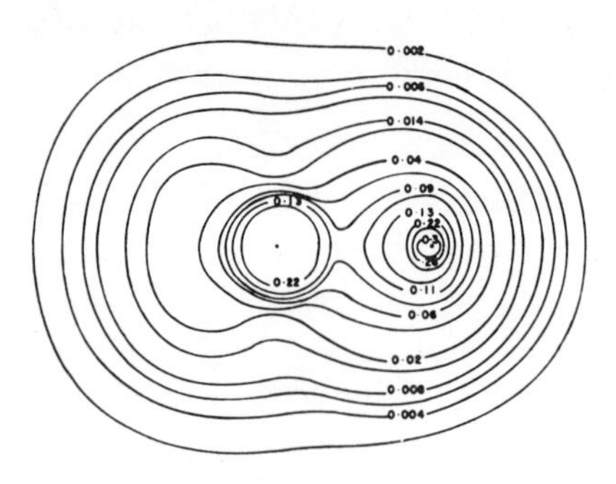
0·002
0·006
0·014
0·04
0·09
0·13
0·22
0·11
0·06
0·02
0·006
0·004
BeH ²Σ⁺

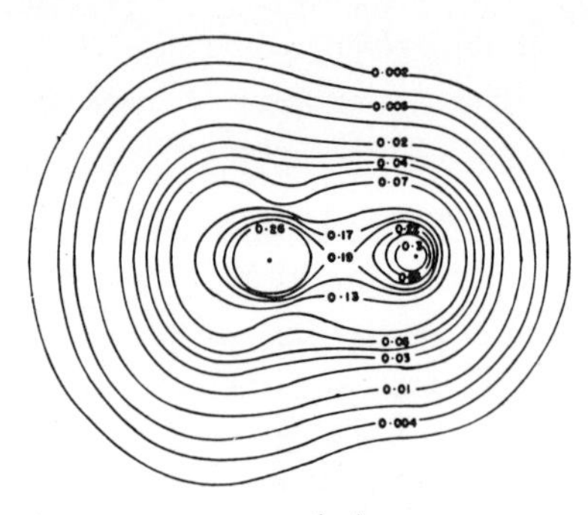
0·002
0·006
0·02
0·04
0·07
0·26
0·17
0·19
0·13
0·03
0·01
0·004
BH ¹Σ⁺

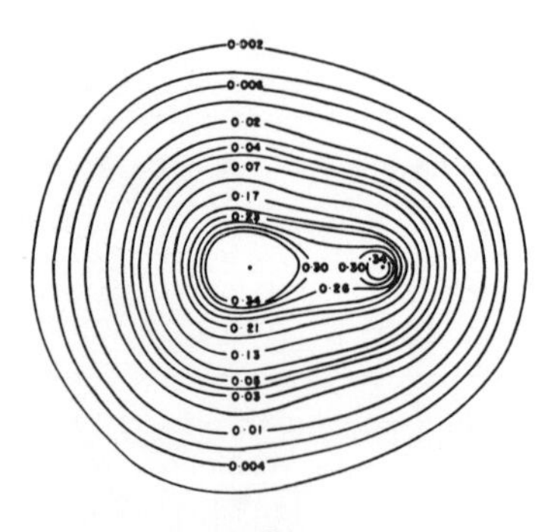
0·002
0·006
0·02
0·04
0·07
0·17
0·30
0·26
0·34
0·21
0·13
0·03
0·01
0·004
CH ²Π_r

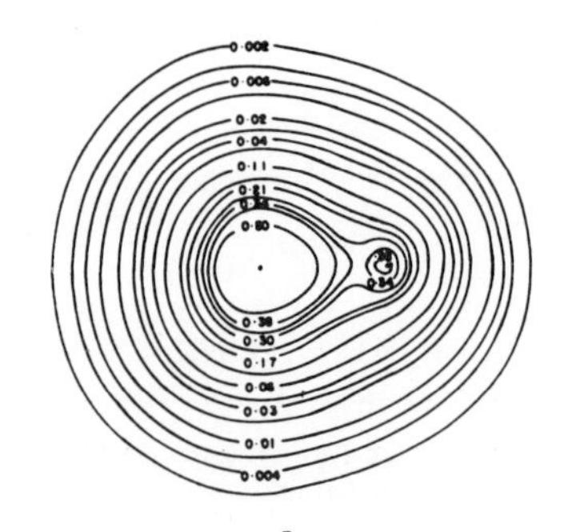
0·002
0·006
0·02
0·04
0·11
0·21
0·30
0·17
0·03
0·01
0·004
NH ³Σ⁻

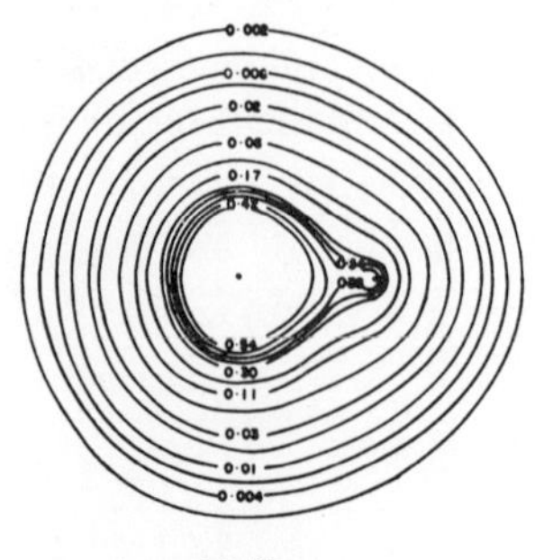
0·002
0·006
0·02
0·17
0·54
0·30
0·11
0·03
0·01
0·004
OH ²Π_I

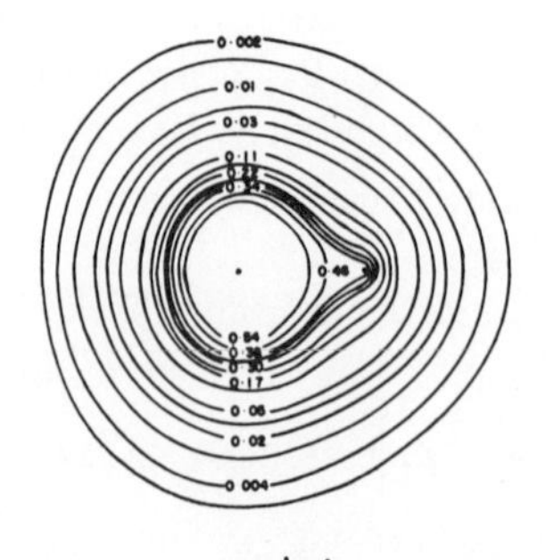
0·002
0·01
0·03
0·11
0·46
0·54
0·30
0·17
0·02
0·004
HF ¹Σ⁺

FIG. 2

little affected by electron correlation (see near end of Section D), the densities based on SCF wave functions may be accepted as good mappings of molecular charge distributions.

An interesting feature [12] of the charge density in LiH is a considerable backward polarization (away from the H) of the Li K shell, attributable to the influence of the H^- on the Li^+. (The structure approximates to Li^+H^-.) LiF shows the same effect.

Bader and Beddall have discussed a "virial partitioning" of charge distributions and energies. The partitioning surface is defined by the path of the gradient vector of the charge density, passing through the point of minimum density between the nuclei [13a].[4] The kinetic and potential energies of molecular fragments determined by such a surface are well defined, and satisfy the virial theorem.

C. POPULATION ANALYSIS AND BONDING

Karo has reported extensively [14] on electron population analysis of LiH and HF, but it seemed better to use Cade and Huo's more accurate SCF wave functions, and from them we have now computed individual MO and total overlap populations (Tables 5–8) and gross atomic populations (Tables 10–13) for the molecules in Tables 1–4. As discussed in Section II.F, the results, while instructive, are somewhat lacking in quantitative significance, especially for the gross populations and the resulting atomic charges (see below).

TABLE 5

Overlap Populations for LiH[a]

$MO(\phi_i)$	$n(i; s_{Li}, H)$	$n(i; p\sigma_{Li}, H)$	$n(i; d, f\sigma_{Li}, H)$	$n(i; Li, H)$
1σ	0.005	−0.000	−0.000	0.005
2σ	0.393	0.334	0.047	0.773
Totals:	0.398	0.334	0.047	0.778

[a] $n(i; s_{Li}, H)$ is the overlap population for ϕ_i summed over all the overlaps of s STFs of Li with all STFs of H, and so on; $n(i; Li, H)$ is the total overlap population for ϕ_i in LiH.

[4] On diatomic hydrides, see Bader and co-workers [13b].

FIG. 2 Total charge densities in diatomic hydrides. [From R. F. W. Bader, I. Keaveny, and P. E. Cade, *J. Chem. Phys.* **47**, 3381 (1967).]

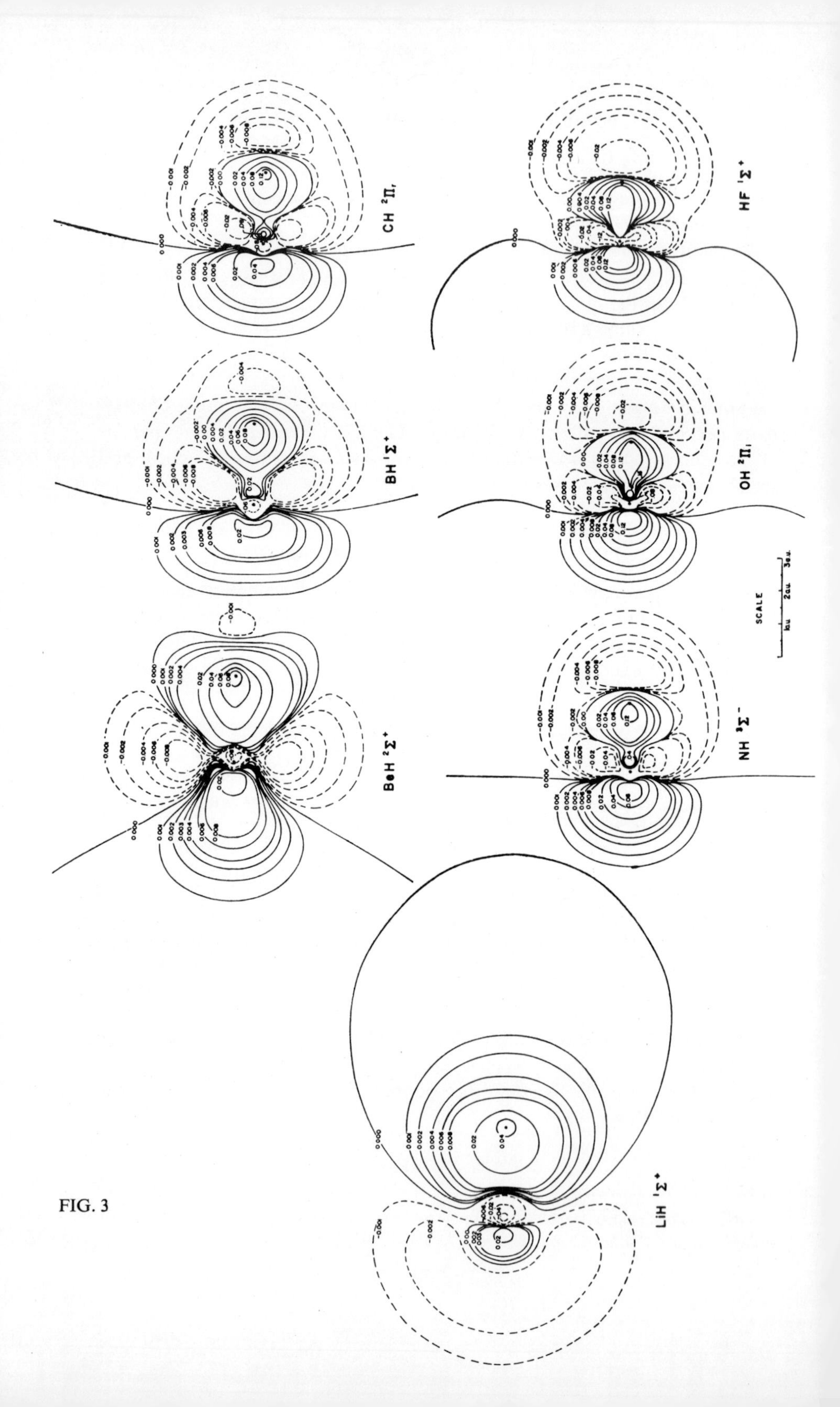
CH $^2\Pi_r$
BH $^1\Sigma^+$
BeH $^2\Sigma^+$
HF $^1\Sigma^+$
OH $^2\Pi_i$
NH $^3\Sigma^-$
LiH $^1\Sigma^+$
SCALE
1a.u. 2a.u. 3a.u.

FIG. 3

The overlap populations n_i in the various MOs can be taken as approximate measures of their contributions to bond strengths and the total overlap population n for a molecule as a measure of the total bond strength [15]. More in detail, partial overlap integrals such as $n(\mathrm{s})$, $n(\mathrm{p}\sigma)$, $n(\sigma)$, and $n(\pi)$ can be taken as measures of the respective contributions of s, pσ, total σ, and π electrons to the bonding.

Probably the best *empirical* measure of bond strength is the dissociation energy D (either D_0, measured from the lowest vibrational level $v = 0$, or D_e, measured from the bottom of the potential curve). Another measure is R_e, which for ground states is usually well correlated with D, but much less so for excited states [16].

A better theoretical measure than n of bond strength D is the product of n times an average ionization potential $\bar{I}$ of the two atoms involved [15]:

$$D = Cn\bar{I}. \tag{1}$$

That a factor such as $\bar{I}$ is needed can be seen by comparing the D values of H_2 and Li_2, which have similar values of n. For H_2, $D_0 = 4.48$ eV and $\bar{I} = 13.60$ eV, which for Li_2, $D_0 = 1.10$ and $\bar{I} = 5.39$ eV. When $\bar{I}$ is reduced, the whole scale of energies, including D, is reduced. Table 9, using computations based on Cade and Huo's tables for first-row hydrides and for HCl, surveys n, $n\bar{I}$, D, and R_e values for these hydrides. A point of interest is the indication from the n_π values that there is some π bonding in the hydrides which contain π electrons, notably in HF and especially HCl.

From the coefficients in Tables 1–4, and even more (see below) from the gross atomic populations in Tables 10–13, it can be seen that there is a considerable amount of s, pσ hybridization in the 2σ and 3σ MOs in the first-row hydrides. (Similarly for the 4σ and 5σ MOs in the second-row hydrides.) This is particularly strong in LiH, where only 2σ is present. The hybridization is positive, that is of the form s+pσ, in 2σ, but more or less negative (s−pσ) in 3σ. As a result, indicated strikingly for CH by the overlap populations for 2σ and 3σ in Table 6, much of the bonding power resides in the pair of electrons in 2σ, in spite of the fact that 2σ is mostly 2s. (In looking at the coefficients in Tables 1–4, keep in mind that it is their *squares*, modified somewhat by overlap terms, which measure their importance—shown directly by the gross atomic populations in Tables 10–13.) In HF, bonding has shifted strongly toward 3σ, but 2σ still shares appreciably as judged by the overlap population. In the series of hydrides from BH to HF, there is a steady shift in the relative overlap populations and the deduced bonding powers of 2σ and 3σ. HCl is similar to HF in that 4σ and 5σ both participate

---

FIG. 3 Difference charge densities compared with atoms for diatomic hydrides. [From R. F. W. Bader, I. Keaveny, and P. E. Cade, *J. Chem. Phys.* **47**, 3381 (1967).]

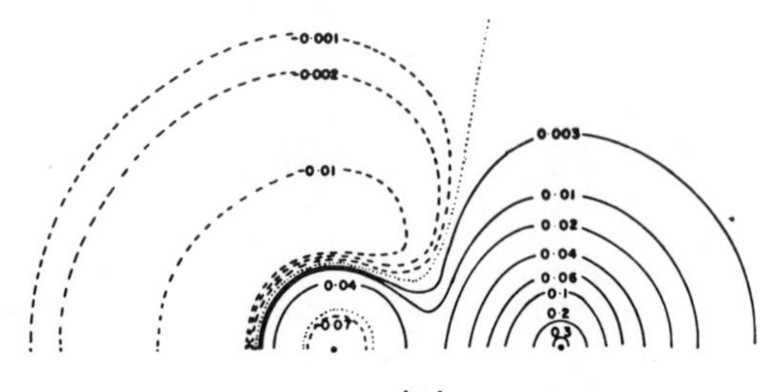

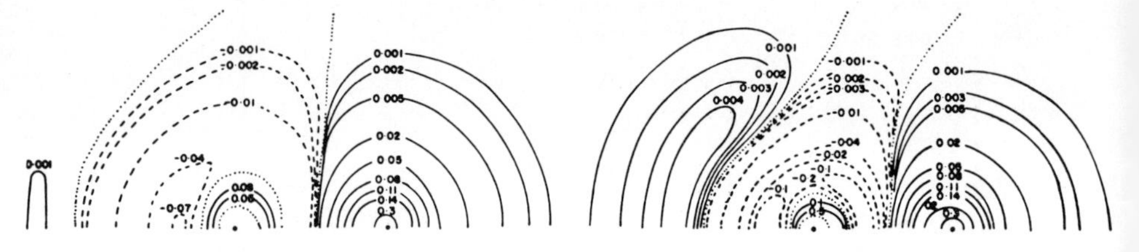

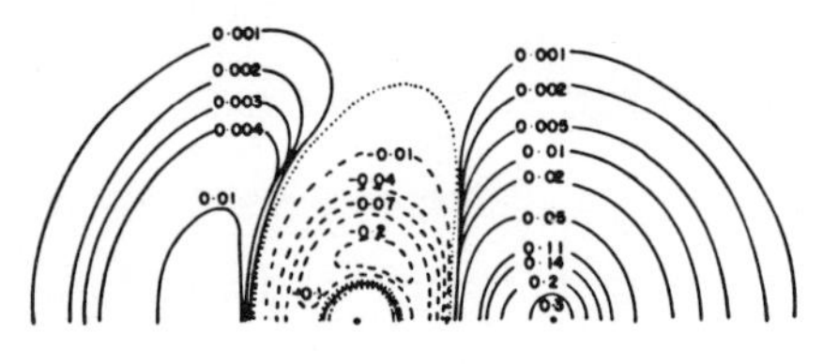

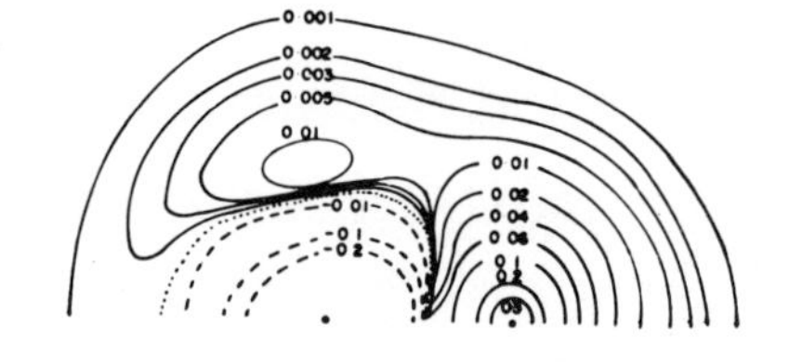

OH ²Π_i

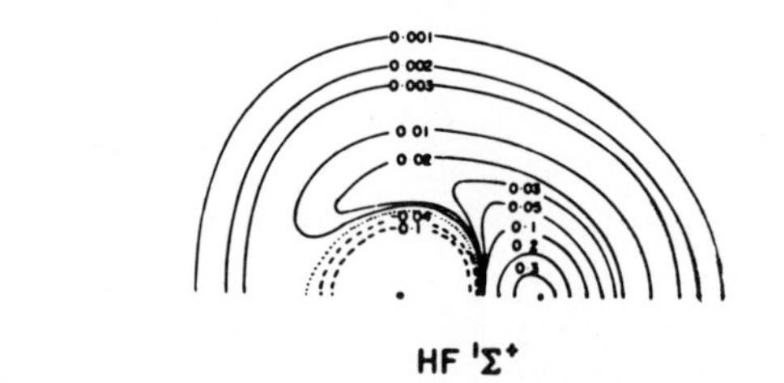

FIG. 4 Difference charge densities compared with united atoms for diatomic hydrides. [From R. F. W. Bader, I. Keaveny, and P. E. Cade, *J. Chem. Phys.* **47**, 3381 (1967).]

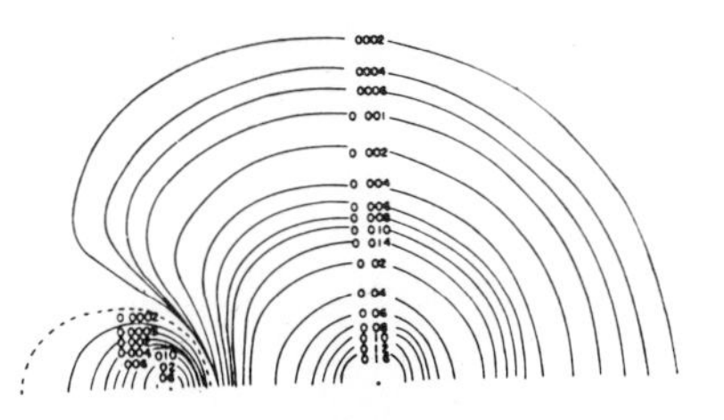

LiH 2σ

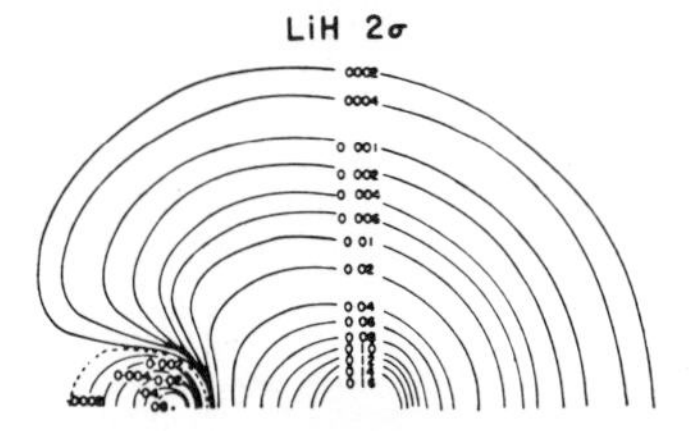

BeH 2σ

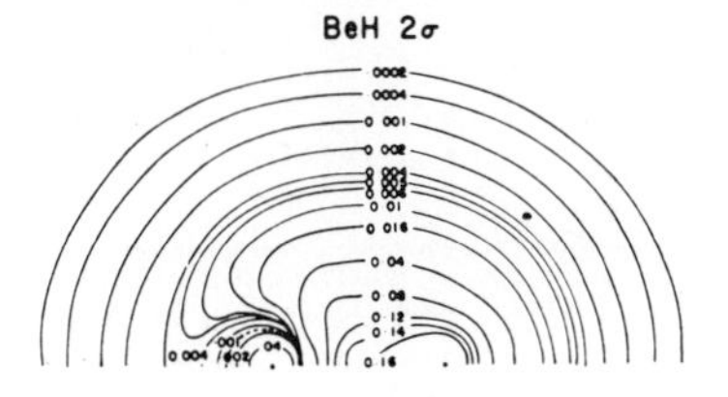

BH 2σ

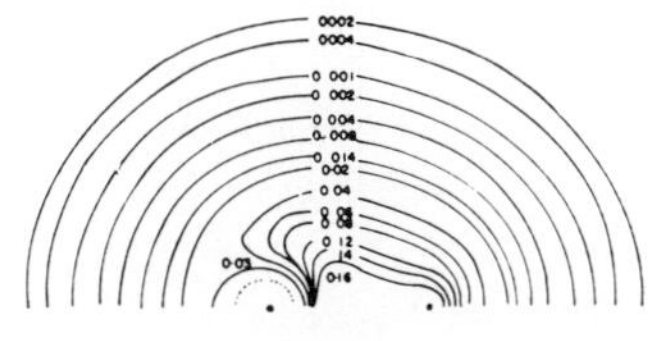

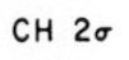

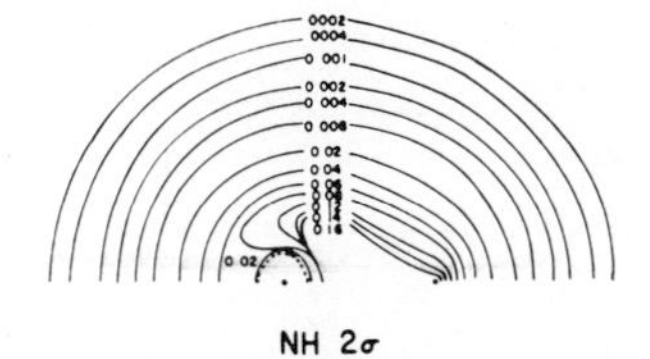

NH 2σ

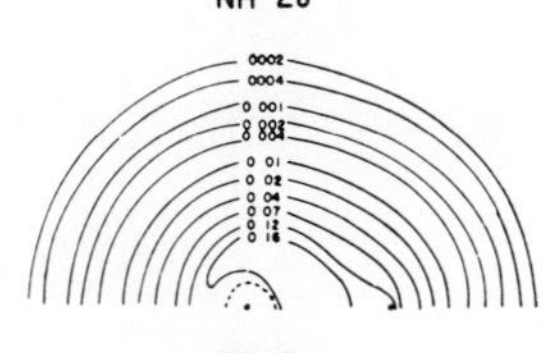

OH 2σ

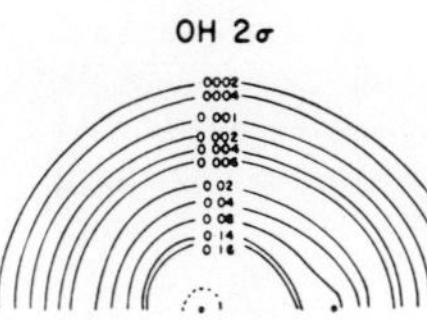

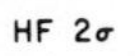

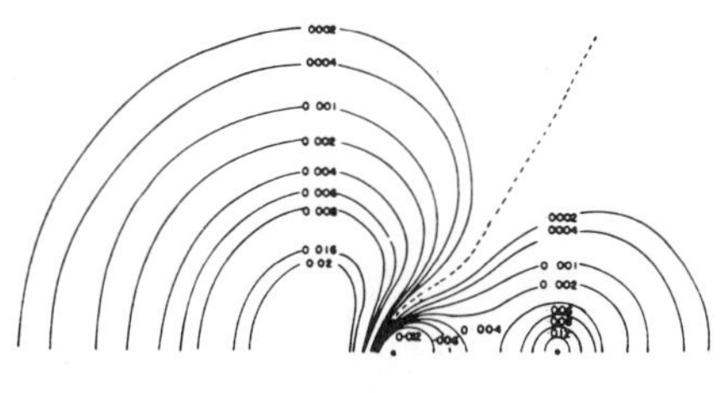

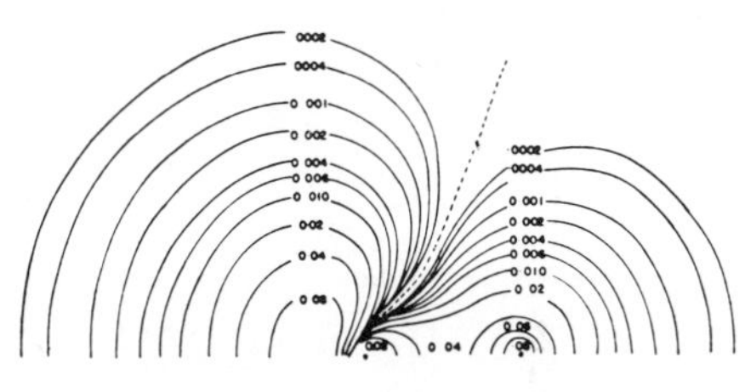

BH 3σ

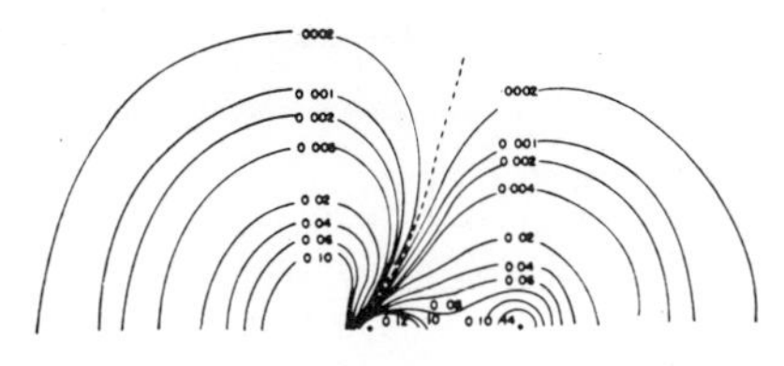

CH 3σ

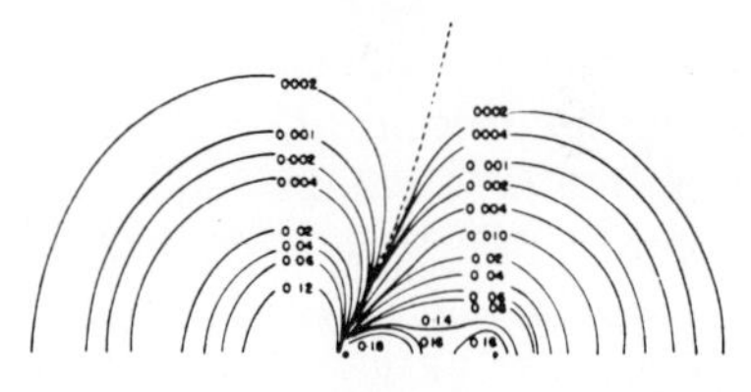

NH 3σ

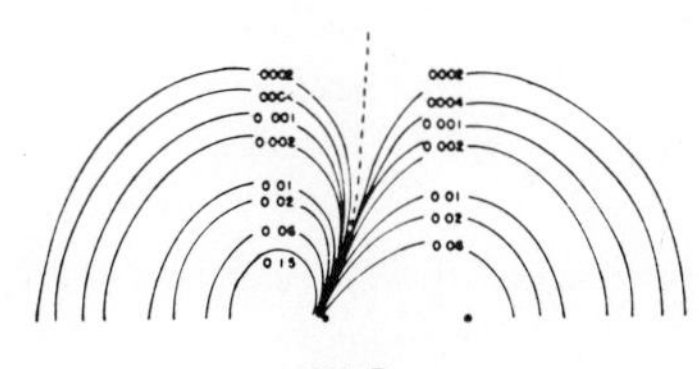

OH 3σ

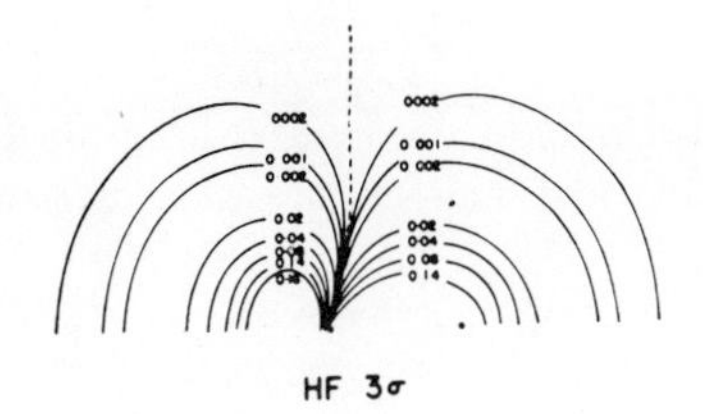

HF 3σ

FIG. 5 Charge densities of 2σ and 3σ MOs for diatomic hydrides. [From R. F. W. Bader, I. Keaveny, and P. E. Cade, *J. Chem. Phys*. **47**, 3381 (1967).]

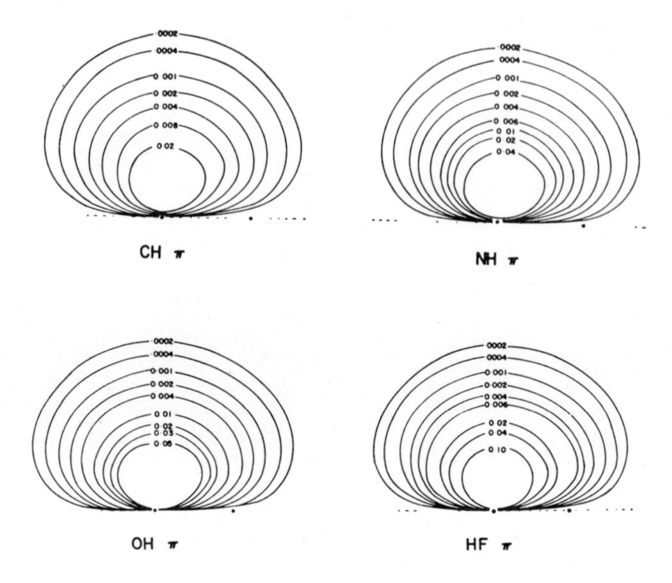

FIG. 6 Contour maps of the 1π molecular orbital charge densities for the first-row hydrides. [From R. F. W. Bader, I. Keaveny, and P. E. Cade, *J. Chem. Phys.* **47**, 3381 (1967).]

strongly in the bonding. The foregoing results contrast with what one would have expected from the usual valence-bond theory, where $2\sigma^2$ would have been identified essentially with a $2s^2$ lone pair and the bonding would have been attributed to $1s_H - p\sigma_A$ bonding in $3\sigma^2$. However, when one looks at the *totals* in the columns headed $n(i; s_A, H)$ and $n(i; p\sigma_A, H)$, these indicate that in the *overall* picture the bonding is essentially just $1s_H - p\sigma_A$. Moreover, when the MCSCF method is used to introduce electron correlation (see Section D), the 2σ MCSCF MOs differ strongly from the 2σ SCF MOs, being antibonding where the SCF MOs are bonding. This can be seen for OH by inspection of Fig. 7 in Section E.4. That the overlap population for the 2σ MCSCF MO of OH is negative is clear also from an examination of its SCF coefficients in Table VI of Ref. 58. On the other hand, the 3σ MO becomes more bonding in the MCSCF wave function. It seems reasonably sure that similar changes of character occur for 2σ and 3σ MOs of the other hydrides on going from the SCF to an MCSCF approximation. In view of the superiority of the latter, our conception of the relative bonding powers of the 2σ and 3σ MOs should probably be guided by the MCSCF MOs. The matter is rather academic, however, since what really counts are the overall charge densities.

From the total gross populations in the columns headed $N(i; s_A)$ and

TABLE 6

Overlap Populations for CH[a]

ϕ_i	$n(i; s_C, H)$	$n(i; p\sigma_C, H)$	$n(i; d, f\sigma_C, H)$	$n(i; p\pi_C, H)$	$n(i; d, f\pi_C, H)$	$n(i; \sigma)$	$n(i; \pi)$	$n(i; C, H)$
1σ	0.001	−0.000	−0.000			0.001		0.001
2σ	0.359	0.117	0.027			0.503		0.503
3σ	−0.359	0.470	0.044			0.155		0.155
1π				0.030	0.001		0.031	0.031
Totals:	0.001	0.588	0.071	0.030	0.001	0.659	0.031	0.690

[a] Cf. footnote to Table 5. But here $n(i; \sigma)$ and $n(i; \pi)$ are the respective total σ and π overlaps for ϕ_i, and $n(i; C, H)$ is the total overlap for ϕ_i.

TABLE 7

Overlap Populations for HF[a]

ϕ_i	$n(i; s_F, H)$	$n(i, p\sigma_F, H)$	$n(i; d, f\sigma_F, H)$	$n(i; p\pi_F, H)$	$n(i; df\pi_F, H)$	$n(i; \sigma)$	$n(i; \pi)$	$n(i; H, F)$
1σ	0.000	−0.000	−0.000			0.000		0.000
2σ	0.108	0.018	0.005			0.130		0.130
3σ	−0.135	0.405	0.027			0.296		0.296
1π				0.085	0.003		0.088	0.088
Totals:	−0.027	0.422	0.031	0.085	0.003	0.426	0.088	0.514

[a] Cf. footnote to Table 6.

TABLE 8

Overlap Populations for Valence Shells of HCl[a]

ϕ_i	$n(i; s_{Cl}, H)$	$n(i; p\sigma_{Cl}, H)$	$n(i; d, f\sigma_{Cl}, H)$	$n(i; p\pi_{Cl}, H)$	$n(i; d\pi_{Cl}, H)$	$n(i; \sigma)$	$n(i; \pi)$	$n(i; H, Cl)$
3σ	0.000	−0.000	−0.000			−0.000		−0.000
4σ	0.275	0.022	0.005			0.302		0.302
5σ	−0.340	0.624	0.035			0.320		0.320
2π				0.199	0.001		0.200	0.200
Totals:	−0.065	0.647	0.041	0.199	0.001	0.622	0.200	0.822

[a] Cf. footnote to Table 6.

TABLE 9

Overlap and Bonding in First-Row Hydrides and HCl[a, b]

Molecule	Calculated n	Observed R_e (a.u.)	Observed $\bar{I}$ (eV)	$n\bar{I}$ (eV)	Observed D_e (eV)
LiH	0.778	3.015	9.5	7.39	2.32
BeH	0.842	2.538	11.5	9.68	(2.6)
BH	0.766	2.336	11.0	8.43	3.56
CH	0.690	2.124	12.5	8.62	3.65
NH	0.606	1.961	14.0	8.48	(3.80)
OH	0.540	1.834	14.5	7.83	4.63
HF	0.514	1.733	15.8	8.12	6.12
HCl	0.822	2.409	13.5	11.10	4.62

[a] For the D_e values, cf. Ref. 1, Table 1 of both papers.
[b] The $\bar{I}$ values are the averages of the ionization potentials of the A and H atoms. They are only rough.

$N(i; p\sigma_A)$ in Tables 10–13, one can obtain the degree of s, pσ hydridization (or s–pσ promotion) in the heavier atom A in these molecules. The extent of promotion, or promotion index, is simply $N(p\sigma_A)$ minus the number of pσ electrons in the free atom; the latter is equal to the total number of p electrons in the free atom, minus the number of π electrons in the molecule. For example in HF, the promotion index is $1.623 - 1 = 0.623$. In addition to s, pσ promotion, there is also a small amount of promotion (one cannot well say whether from s or p) to dσ and fσ, also from pπ to dπ, in the molecule. One could also look at s $\rightarrow$ pσ and pπ promotion in the H atom in each molecule, but it has not seemed worthwhile to include this detail in the tables, except for the π promotion indicated under $N(i; \pi_H)$.

Symbolically, one can conveniently summarize the gross population in

TABLE 10

Gross Atomic Populations for LiH[a]

ϕ_i	$N(i; s_{Li})$	$N(i; p\sigma_{Li})$	$N(i; d, f\sigma_{Li})$	$N(i; \sigma_{Li})$	$N(i; H)$
1σ	1.997	0.000	−0.000	1.997	0.003
2σ	0.384	0.261	0.026	0.672	1.328
Totals	2.381	0.261	0.026	2.669	1.331

[a] $N(i; s_{Li})$, $N(i; p\sigma_{Li})$, and $N(i; d, f\sigma_{Li})$ are the partial gross populations in s_{Li}, $p\sigma_{Li}$, and $d\sigma_{Li}$ plus $f\sigma_{Li}$ in ϕ_i; $N(i; H)$ is the partial gross population of the basis STFs of H.

TABLE 11

Gross Atomic Populations for CH[a]

ϕ_i	$N(i; s_C)$	$N(i; p\sigma_C)$	$N(i; d, f\sigma_C)$	$N(i; \sigma_C)$	$N(i; \sigma_H)$	$N(i; p\pi_C)$	$N(i; \pi_C)$	$N(i; \pi_H)$	$N(i; H)$
1σ	2.000	0.000	−0.000	2.000	0.000				0.000
2σ	1.432	0.154	0.019	1.605	0.395				0.395
3σ	0.461	1.222	0.034	1.716	0.284				0.284
1π						0.982	0.983	0.017	0.017
Totals:	3.893	1.376	0.052	5.321	0.679	0.982	0.983	0.017	0.696

[a] Cf. footnote to Table 10.

TABLE 12

Gross Atomic Populations for HF[a]

Φ_i	$N(i; s_F)$	$N(i; p\sigma_F)$	$N(i; d, f\sigma_F)$	$N(i; \sigma_F)$	$N(i; \sigma_H)$	$N(i; p\pi_F)$	$N(i; \pi_F)$	$N(i; \pi_H)$	$N(i; H)$
1σ	2.000	0.000	0.000	2.000	0.000				0.000
2σ	1.884	0.032	0.003	1.919	0.081				0.081
3σ	0.091	1.592	0.019	1.702	0.298				0.298
1π						3.949	3.953	0.047	0.047
Totals:	3.975	1.623	0.023	5.621	0.379	3.949	3.953	0.947	0.426

[a] Cf. footnote to Table 10.

TABLE 13

Gross Atomic Populations for HCl[a]

Φ_i	$N(i; s_{Cl})$	$N(i; p\sigma_{Cl})$	$N(i; d\sigma_{Cl})$	$N(i; \sigma_{Cl})$	$N(i; \sigma_H)$	$N(i; p\pi_{Cl})$	$N(i; \pi_{Cl})$	$N(i; \pi_H)$	$N(i; H)$
3σ	0.000	2.000	0.000	2.000	−0.000				−0.000
4σ	1.752	0.020	0.004	1.776	0.224				0.224
5σ	0.156	1.226	0.027	1.409	0.591				0.591
2π						3.890	3.891	0.109	0.109
Totals:	1.908	3.246	0.030	5.185	0.815	3.890	3.891	0.109	0.924

[a] Cf. footnote to Table 10.

any molecule by means of configuration symbols giving the electronic occupation number for each type of STF; for example, for LiH,

$$s_{Li}^{2.38}\, p_{Li}^{0.26}\, d_{Li}^{0.02}\, s_{H}^{1.33}$$

(cf. Table 10), and for CH,

$$s_{C}^{3.89}\, p_{C}^{2.36}\, d_{C}^{0.05}\, s_{H}^{0.67}\, p_{H}^{0.03},$$

or in more detail, CH,

$$s_{C}^{3.89}\, p\sigma_{C}^{1.38}\, d\sigma_{C}^{0.05}\, f\sigma_{C}^{0.004}\, p\pi_{C}^{0.96}\, d\pi_{C}^{0.02}\, s_{H}^{0.67}\, p\sigma_{H}^{0.01}\, p\pi_{H}^{0.02}.$$

The computed charges, in electron units, on the atoms can be obtained as follows: for the H atom, subtract 1 (the number of electrons on the free H atom) from the total $N(\mathrm{H})$ in the column headed $N(i;\,\mathrm{H})$. The computed charge on the A atom is of course just the negative of this. Table 14 shows

TABLE 14

Calculated H Atom Charges for First-Row Hydrides and HCl

Molecule	Q (electrons) Pop. analysis	Q (electrons) Politzer	Molecule	Q (electrons) Pop. analysis	Q (electrons) Politzer
LiH	−0.331	−0.36	NH	0.425	0.10
BeH	−0.152	−0.14	OH	0.518	0.19
BH	0.000	−0.05	HF	0.574	0.27
CH	0.304	0.03	HCl	0.077	

the values of Q_{H} here obtained from Cade and Huo's results for the first-row hydrides and HCl, and also shows values obtained by Politzer by a different method (see below).

As was indicated in Section II.E, however, we must not attach precise significance to the computed atomic charges of Table 14. Kern and Karplus [17] have compared the charge distributions and computed Q values for five of the earlier SCF computations on HF—all less accurate than those of Cade and Huo. From the results of Nesbet [18] (18 basis functions, $E = -100.0571$ a.u.; cf. Cade and Huo, 24 functions, $E = -100.0703$ a.u.), they compute $Q = +0.23$e. On the other hand, from an SCF function of Clementi [19] (16 basis functions, $E = -100.0575$ a.u.), they compute $Q = +0.48$e. Yet charge-density maps show only a slight difference between Nesbet's and Clementi's functions; further, this near-equality is confirmed in an analysis by Politzer [20]. Finally, our computed Q from Cade and Huo's HF function is +0.574e.

The problem of the "true" charges on atoms in a molecule is a difficult one, and in spite of many attempts,[5] no satisfactory solution seems to have been found, if indeed there is one. However, a recent discussion [21b] in terms of MAOs [cf. Eq. (II.3)] may be promisnig.

As Mulliken pointed out [22], a good molecular wave function can be constructed even from basis functions which are centered on just one atom. The values of physical observables computed with such a wave function can be quite accurate; yet a population analysis will assign *all* the electronic charge to just that one atom. It was concluded that if meaningful atomic charges are to be obtained from a population analysis, the wave functions must be built up from "balanced" basis sets. Unfortunately there does not seem to be any good way to ensure in the case of heteropolar molecules that a basis set is indeed balanced. One must use judgment, and this is fallible. Roughly speaking, one might look for a basis set with maximally rapid convergence.

Politzer, however, has proposed a method which looks promising, at least for diatomic and linear molecules [23]. It involves integrating the electronic density function over regions associated with the individual atoms, the regions being defined in terms of the superposed charge distributions of the corresponding free atoms. When this method is used, Nesbet's function for HF gives $Q_H = +0.27$e, while Clementi's gives +0.26e. Table 14 also shows Politzer's results[2] for most of the other hydrides, using Cade and Huo's wave functions. In the heavier hydrides the results are notably smaller than those from our population analysis. However, Politzer's result[6] for LiF ($Q_{Li} = +0.52$) seems too small, and one wonders whether somewhat larger values for Q_H for these hydrides would not be better.

D. ELECTRON CORRELATION

Except for the lightest molecules (1- to 4-electron diatomic molecules), where the James–Coolidge method can be used, with its explicit introduction of r_{12}, electron correlation is usually obtained by expressing the wave function as a sum of terms, each corresponding to a different electronic configuration (cf. Section I.D):

$$\Psi = \sum_i C_i \Phi_i . \tag{2}$$

[5] See citations in Ref. 20; see also Roby [21a].

[6] The data are from a private communication from Professor Politzer.

In practice, it is necessary to confine oneself to a finite number of terms; experience shows that even with the thousands of terms that have sometimes been used, one is still some distance from the limit of an accurate Ψ.

When Eq. (2) is used, it is necessary to extend the definitions of population analysis already given. Since the Φ_i's in Eq. (2) are orthogonal, this can be done in an obvious way by computing the desired partial or total populations in each Φ_i of Eq. (2), and adding the results [24].

In most molecules at R_e, there is one *dominant* electron configuration Φ_1 ($C_1 \approx 0.98$ or so). In closed-shell cases, this corresponds to a single Slater determinant, but in general a configuration state function (CSF) which is a linear combination of Slater determinants is needed to represent Φ_1 and other Φ's. Ordinary SCF calculations approximate Ψ by Φ_1, which is ordinarily called *the* electron configuration. In a few systems, more than one configuration is important even at R_e.

At large R values, at least two and often more Φ's are absolutely necessary for most molecules for a decent approximation to Ψ. Exceptions are cases such as the lowest $^3\Sigma^+$ MO of H_2 or the ground state of a system built from two closed-shell partners (e.g., He_2). Here an SCF function with its single dominant MO configuration remains a good approximation over all or at least a large range of R values; however, the detailed forms of some of the MOs involved sometimes change radically with R (cf. Section E.2 for an example in BH).

In the early work on electron correlation, one usually began with an SCF calculation using a set of STFs as basis functions. Given M basis functions, for a molecule whose dominant configuration contains m different occupied MOs, one found m solutions for the occupied MOs, plus $M-m$ virtual MOs that are orthogonal to the occupied MOs. Using all the M occupied and virtual MOs, one can set up a large number of "excited" Φ's by populating these with varying numbers of electrons adding up to the total number in the molecule.

To illustrate, we give in Table 15 the results of an early calculation by Fraga and Ransil [25][7] on LiH using a minimal basis set. Here $M = 4$, for a basis set consisting of $1s_{Li}(2.691)$, $2s_{Li}(0.708)$, $2p\sigma_{Li}(0.845)$, and $1s_H(0.977)$, and $m = 2$. By redistributing the four electrons in all possible ways using the total of M MOs, one finds thirteen different Φ's, called I to XIII in Table 15. Table 15 gives the coefficient C_i for each Φ_i of Eq. (2). Note that all the coefficients are small compared with that of the dominant configuration; note also that most of the coefficients are negative, as is typical. An interesting exceptional case where there are two important configurations is

[7] This paper contains calculations on a number of homopolar and heteropolar molecules containing H and/or first-row atoms.

TABLE 15

Coefficients for a Minimal-Basis LiH Wave Function with Limited CM

Configuration	Electron distribution				C_i	CM wave function
	MO: 1σ	2σ	3σ	4σ		
I	2	2			C_{I}	0.989
II	2		2		C_{II}	−0.053
III	2			2	C_{III}	−0.093
IV		2	2		C_{IV}	−0.003
V		2		2	C_{V}	−0.002
VI	2		1	1	C_{VI}	0.089
VII		2	1	1	C_{VII}	−0.001
VIII	1	1	2		C_{VIII}	−0.001
IX	1	1		2	C_{IX}	−0.001
X	2	1	1		C_{X}	0.055
XI	2	1		1	C_{XI}	−0.004
XII	1	2	1		C_{XII}	−0.000
XIII	1	2		1	C_{XIII}	0.000

that of the ${}^1\Sigma_g^+$ ground state of C_2, where with a minimal basis set c_{I} (for $1\sigma_g^2 1\sigma_u^2 2\sigma_g^2 2\sigma_u^2 1\pi_u^4$) is 0.896 and c_{II} (for $1\sigma_g^2 1\sigma_u^2 2\sigma_g^2 1\pi_u^4 3\sigma_g^2$) is −0.426 [25].

When an extended basis set is used, as is necessary for an accurate wave function, a huge number of Φ's is obtained. From this number, one endeavors to select the most important. The process using virtual MOs is inefficient, however, mainly because in the excited Φ's these MOs are of large size, whereas for rapid convergence one should use Φ's built from MOs not very different in size from those in the dominant configuration.

Systematically, the problem can be solved to a considerable extent by using the MCSCF method in which the optimal MOs for two or a few of the most obviously important Φ's in Eq. (2) are obtained by a generalized SCF procedure. These Φ's are chosen in such a way that as R increases, Ψ dissociates to the correct pair of atomic states. Additional Φ's can also be included in limited number. At this point one can transform to configurations built from NOs (natural orbitals) which are a set of MOs so chosen as to give maximum speed of convergence in the series indicated in Eq. (2) (see Section I.F). Many recent calculations make use of NOs, with a preceding CM or MCSCF calculation. The MCSCF MOs are already approximate NOs. In the MCSCF procedure, even valence-shell MOs determined thereby often differ to an important extent from their like-named

SCF counterparts [see second and third paragraphs after Eq. (1) in Section C for a discussion and Section E.4 for examples].

For the ground states of the first-row hydrides at their R_e's, Bender and Davidson [26] have obtained Eq. (2) Ψ's yielding 70% or more of the correlation energy, using approximate NOs (see Section I.F). They used basis sets for the NOs which consisted of 17σ, 17π, 6δ, and 1ϕ STFs, and from 939 to 3379 CSFs for the various molecules. They included Φ's which correlate not only the valence shell MOs but also the K-shell (1σ) MOs.

In earlier work [27] on LiH, they showed (see Table 16) that although

TABLE 16

The Most Important Configurations from the Bender–Davidson LiH Calculation

Configuration	C_i^2	E_k
1. $1\sigma^2 2\sigma^2$	0.9718	−7.9871
2. $1\sigma^2 3\sigma^2$	0.0114	−8.0023
3. $2\sigma^2 2\pi^2$	0.0010	−8.0161
4. $1\sigma^2 1\pi^2$	0.0104	−8.0289
5. $2\sigma^2 5\sigma^2$	0.0008	−8.0391
6. $2\sigma^2 6\sigma^2$	0.0005	−8.0451
7. $1\sigma^2 4\sigma^2$	0.0029	−8.0494
8. $2\sigma^2 4\sigma 5\sigma$	0.0002	−8.0514

[a] The energy E_k associated with the kth configuration is that obtained from the CM including all configurations through the kth.

the *coffiecients* for the 1σ correlations (Nos. 3, 5, 6, 8) are small, their contributions to the correlation *energy* are disproportionately important. This is evidently related to the fact that the 1σ shell contributes most of the energy. In most calculations to date, however, the K-shell (or in general inner-shell) correlation energy is omitted. This is to a large extent justified if one is interested, as is usual, mainly in a curve showing how the energy varies when atoms come together. For one can reasonably assume that the inner electrons and their correlation energy remain about the same in the molecule as in the separate atoms (but see contrary indications mentioned in Section E.1).

From their Ψ's for LiH, Bender and Davidson [26] computed the expectation values of numerous functions of the coordinates, including charge density, spin density, field gradient, and force on the nucleus at each of the two centers A and H in an AH molecule; also the dipole moments.

In a series of papers, Krauss, Stevens *et al.* [28] have determined MCSCF wave functions for CH, NH, OH, and HF by the OVC method, using eight configurations. Other examples of the application of the MCSCF approach are discussed in Section E.

The principal aim of the OVC method as first proposed by Das and Wahl was to add a minimal number of CM wave functions to the SCF function such as to assure that the energy of the total wave function as $R \to \infty$ would become equal to the *SCF energy* of the separate atoms into which the molecule in the particular state concerned dissociates. In this way, by subtracting the OVC energy from the SCF energy, of the separate atoms, the true dissociation energy of the molecule would be fairly well approximated.

For a really accurate dissociation energy, however, further CM terms must be added. The OVC or MCSCF method *in general* includes the most important of these terms. Without being exhaustive as to energy minimizing CM, it seeks to include those CM terms which change with R enough to affect calculated molecular properties appreciably, and most especially emphasizes those necessary to give a good potential curve.

The excess of the dissociation energy over the SCF-computed dissociation energy into SCF atoms is called the "molecular extra correlation energy" (MECE). OVC methods seek to approximate this fairly closely.

Every MCSCF calculation begins with the setting up of a *reference function* or base function containing as a minimum enough Φ's to ensure a fairly good approximation for all R values out to dissociation. The dominant Φ at R_e may still be present as $R \to \infty$ or it may vanish, while in general one or more additional Φ's become necessary at large R values. Thus, for example, for the $^2\Pi$ ground state of OH, the dominant Φ at R_e is of the configuration $1\sigma^2 2\sigma^2 3\sigma^2 1\pi^3$, but as R increases $1\sigma^2 2\sigma^2 4\sigma^2 1\pi^3$ and $1\sigma^2 2\sigma^2 (3\sigma 4\sigma, {}^3\Sigma^+) 1\pi^3$ must be added. Here 3σ and 4σ are respectively of the approximate LCAO forms $a2\text{p}\sigma_0 + b1\text{s}_\text{H}$ and $a1\text{s}_\text{H} - b2\text{p}\sigma_0$ $(a > b)$. The MOs included in the reference function have been called *internal* MOs; these in turn are divided into *core* and *valence-shell* MOs. Higher-energy MOs have been called *external* orbitals.[8]

The OH valence function just cited may be called a minimal reference function. However, MCSCF calculations are often made with more extended reference functions. For example, one or more external MOs may be used in some Φ's, or often, additional Φ's can be found using internal MOs only; as an example of the latter the $1\sigma^2 2\sigma^2 3\sigma^2 4\sigma^2 1\pi$ configuration may be included, although actually this configuration has been found of little importance. Another possibility is $1\sigma^2 3\sigma^2 4\sigma^2 1\pi^3$.

After a reference function has been chosen, additional configurations

[8] See McLean and Liu [29a] and Bagus *et al.* [29b] for a systematic analysis of CSFs in CM into zero-order, first-order, and so on.

obtained by transferring electrons from internal to external MOs can be introduced to effect a CM calculation. For example, in OH one can add configurations such as $1\sigma^2 2\sigma^2 1\pi^3 (2\pi^2, {}^1\Sigma^+)$ or $1\sigma^2 2\sigma^2 1\pi^3 5\sigma^2$. In general, it is necessary to specify the internal coupling in any open shells which are involved, as here in $\ldots(2\pi^2, {}^1\Sigma^+)$, and above in the OH wave function, $\ldots(3\sigma 4\sigma, {}^3\Sigma^+) 1\pi^3$.

The specific linear combination of Slater determinants which is needed here for a given Φ is known as a CSF (configuration state function). Only in simple special cases does a CSF consist of a single Slater determinant. In some cases different but equivalent choices of CSFs as $R \to \infty$ are available; the best choice is one in which the forms of the individual MOs (which often change radically as $R \to \infty$) change as smoothly as possible.

Krauss *et al.* [30] have classified the various types of Φ's involved in MCSCF and CM somewhat as follows:

(1) single-electron substitutions by internal or external MOs (e.g., $3\sigma \to 4\sigma$, or, $1\pi \to 2\pi$);

(2) single-shell substitutions, either by internal MOs (e.g., $3\sigma^2 \to 4\sigma^2$ in OH) or by external MOs (e.g., $3\sigma^2 \to 2\pi^2, {}^1\Sigma^+$ in OH); these are *double excitations* or *pair excitations*;

(3) split-shell double excitations (e.g., $3\sigma^2 1\pi^3 \to (3\sigma 4\sigma, {}^3\Sigma^+) 1\pi^2 2\pi$ in OH);

(4) higher excitations. Rather than "substitutions," the word "excitations" is commonly used, but "substitutions" is preferable because the Φ's in CM do not correspond to real excited states, but lie within the same geometric space as that occupied by the SCF or the MCSCF reference function; this fact becomes obvious when NOs are used.

As Billingsley and Krauss, with applications to OH and CO, point out [30], some CSFs in an OVC or MCSCF base function effect transfers of charge within the system; for example $3\sigma^2 \to 4\sigma^2$ in OH transfers electrons from an MO(3σ) predominantly on the O atom to one (4σ) predominantly on the H atom. In homopolar molecules, there are no such net charge transfers, but there can be changes in the degree of ionicity (in terms of valence-bond theory). For example in H_2, the partial replacement of $1\sigma_g{}^2$ by $1\sigma_u{}^2$ is equivalent to a reduction in ionicity (i.e., in H^+H^- plus H^-H^+ character).

The dipole moment is a property whose quantum-mechanical operator obviously involves coordinates only, and thus is one of a class of one-electron properties for which the SCF wave function should be correct to second order of perturbation theory [31]. But how good is that? Calculations by Cade and Huo [32][9] show that agreement with experiment is fairly good

[9] First- and second-row hydrides.

TABLE 17

Dipole Moments of First-Row Hydrides (in Debye Units)

	SCF	Eq. (2)	Experiment
LiH	6.002	5.853	5.82
BeH	0.282	0.248	
BH	−1.733	−1.470	
CH	−1.570	−1.427	−1.46±0.06
NH	−1.627	−1.587	
OH	−1.780	−1.633	−1.66
HF	−1.942	−1.816	−1.82

for the SCF wave functions of first-row hydrides, but that the calculated values are about 10% too high. Table 17 compares the computed SCF values, the computed values using Bender and Davidson's correlated wave functions [26], and the experimental values. It is seen that the agreement with experiment is excellent for the correlated wave functions; evidently these include practically all the Φ's that influence the dipole moments appreciably.

Several authors [30] have pointed out that most of the improvement as compared with SCF comes from the inclusion of singly substituted internal Φ's. Green [33a], in connection with a calculation on LiH only slightly less accurate than that of Bender and Davidson, gives a discussion of this point and gives three references to earlier papers. In these papers, it is noted as a *general characteristic of one-electron properties* that for them it is one-electron substitutions that are essentially responsible for the improvements effected by CM. As Green points out, however, this result is true only in the presence of *interaction* with double substitutions; a balance between single and double substitutions is necessary. The single substitutions are coupled through the double substitutions. For further discussions of the accuracy of calculated dipole amounts see Green's thorough review [33b]; see also Sections VI.A and VI.F.

The charge density, like the energy, is a property for which the SCF wave function should give correct results to the second order of perturbation theory [34].[10] Banyard and Hayns [35] have made a careful comparison between Cade and Huo's SCF charge densities for LiH and the very good correlated wave function (giving 89% of the correlation energy) by Bender and Davidson [27]. They find that the charge cloud is slightly expanded (perhaps 1 or 2%) in the correlated as compared with the SCF wave function. Further, the electron density is slightly increased close to each nucleus

[10] Also compare Ref. 34b regarding the charge density.

and slightly reduced in the internuclear region. They give contour diagrams showing these changes and other details. It is seen that the charge density is much less sensitive than the dipole moment to changes produced by electron correlation. In cases where electron correlation is unusually large, however, for example in most molecules at large R values, an ordinary SCF charge density is no longer so good an approximation (cf. Section V.G).

Instead of trying to *compute* correlation energies, an alternative at the present stage is to obtain a semiempirical expression for them. Lie and Clementi, elaborating on work of Gombas, have developed a suitable functional which they have applied to calculations of the correlation energies of the first-row hydrides, not only at R_e, but at R values out to dissociation [36a].[11] In doing so, however, they had in general to start with MCSCF rather than simple SCF functions, so as to get correct dissociation behavior. Agreement with experiment is very good for dissociation energies but otherwise only moderately good.

The discussion thus far has bene largely in terms of straightforward CM methods, in particular OVC and MCSCF methods. Although successful in terms of potential curves and dissociation energies, the slow convergence of these methods does not lend itself to obtaining the major portion of the correlation energy. Moreover, they involve the rather poorly founded assumption, which can often lead to appreciable errors, that intraatomic correlation energies do not change during molecule formation (i.e., the concept of MECE, Section D).

Meyer *et al.* have recently been developing alternative approaches which provide much more rapid convergence toward the full correlation energy, and at the same time provide, within limits, more accurate potential and dipole moment curves as a function of R, although as yet not quite the best dissociation energies. These methods are reviewed briefly in Section I.F. The first step was the PNO–CI method using pseudonatural orbitals (PNO) combined with CM; more than 90% of the total correlation energy was accounted for in a study of the ionization energies of water [37]. This variational method was soon supplemented by improvements using coupled electron pairs (CEPA). Although not strictly variational, the combined PNO–CI and CEPA methods [38] led to some excellent agreements of spectroscopic constants with experiment, particularly for OH [39] and for first- and second-row diatomic hydrides in general [40]. Most recently, Meyer *et al.* have begun the development of a further variational method, that of self-consistent electron pairs (SCEP) [41].

The results of the PNO–CI and CEPA studies of Meyer and Rosmus (MR) on diatomic hydrides will now be reviewed. MR made a systematic

[11] See Ref. 36b for a similar discussion on homopolar diatomic molecules.

study of the ground-state potential curves and dipole moments as functions of R for all the diatomic hydrides LiH to HCl. The computations required only about twice the computing times of conventional SCF calculations, and can be performed just as routinely. The potential curves are extremely good near R_e but not out to free dissociation; errors of the calculated dissociation energies reach 0.3 eV. For the potential curves, however, significant deviations from the experimental curves do not occur for R values less than $2R_e$. In this range near R_e, between 95% (LiH) and 85% (HCl) of the valence-shell correlation energies are accounted for in the CEPA calculations. For LiH and BeH, MR included in the calculation the K-intrashell and the KL-intershell correlations and for NaH, MgH, and AlH, the LM-intershell correlation.

Comparison of the spectroscopic constants derived from the CEPA theoretical curves with experiment shows a high reliability of the theoretical values. The standard deviations over both first- and second-row hydrides are: R_e, 0.003 Å; ω_e, 14 cm^{-1}; α_e, 0.005 cm^{-1}; and $x_e\omega_e$, 1.5 cm^{-1}. Several vibrational-level intervals ΔG (0–1, 1–2, and 2–3) are computed and compared with experiment. MR give detailed tables and figures comparing the results of SCF, PNO–CI, CEPA, previous work, and experiment for the spectroscopic constants of the hydrides. For the lighter hydrides in each row, they also tabulate the respective contributions to the spectroscopic constants and μ_e due to intravalence-shell versus core and core-valence-shell correlations. The latter contributions are important for LiH and NaH, but become rapidly less so for succeeding row members. Various vibrational matrix elements have been calculated from the dipole moment curves, which are presented *in extenso*. The μ_0 values show errors of 0.02 to 0.04 D.

E. CALCULATIONS ON SELECTED MOLECULES

1. LiH

Numerous *ab initio* calculations have been made on LiH [2]. For the ground state, a calculation by Bender and Davidson using elliptical coordinates [27] was until recently the most accurate to date, except for a transcorrelated wave function by Boys and Handy [42]. Far more accurate are the calculations of Meyer and Rosmus, discussed at the end of Section D. Much less accurately, Bender and Davidson [43] reported potential curves, dipole moments, and oscillator strengths for the first 19 states.

Docken and Hinze [44][12] made a more detailed and accurate study by the MCSCF method of the five lowest states. They left the $1\sigma^2$ shell un-

[12] See also the variational time-dependent Hartree–Fock calculations on LiH and BeH^+ by Stewart *et al.* [45].

correlated, with the thought that this should make relatively little difference to the derived properties, in particular the potential curves. On this matter, reference should be made to Section D for relevant calculations of Meyer and Rosmus.

Docken and Hinze used a basis set of 23σ STFs (15 on Li and 8 on H), 8π STFs (5 on Li, 3 on H), and 4δ STFs (2 each on Li and H). They used 15 Φ's for the X $^1\Sigma^+$ (ground) and the A $^1\Sigma^+$ (first excited) state, five Φ's for the B $^1\Pi$ and the related (experimentally unknown) $^3\Pi$ state, and four Φ's for the predicted repulsive $^3\Sigma^+$ state. (Note that letter symbols, such as A in A $^1\Sigma^+$, indicate experimentally known states.) They calculated potential curves ($R = 2$ to 12 a.u.) and corresponding spectroscopic constants for the stable states, also transition moments and line strengths of individual band lines for transitions from the ground to the excited states, and a variety of molecular properties, including dipole and quadrupole moments and field gradients at the nuclei. They give details of the forms of the MOs at several R values.

2. BeH and BH

An SCF calculation of the ground state of BeH has been discussed in Section A. Correlated wave functions have been obtained for this and for the A $^2\Pi$ lowest excited state by Bagus *et al.* [46].

The ground state of BH is $1\sigma^2 2\sigma^2 3\sigma^2$, $^1\Sigma^+$. The lowest two excited states $1\sigma^2 2\sigma^2 3\sigma 4\sigma$, $^3\Sigma^+$ and B $^1\Sigma^+$, are of especial interest. Experimentally, the B $^1\Sigma^+$ state is a Rydberg state having nearly the same R_e as the ground states of BH and BH^+. SCF calculations show that at R_e the $^3\Sigma^+$ state is also a Rydberg state [47], yet as $R \to \infty$ it must become a repulsion state dissociating into ground state B+H [48]. After a minimum at R_e (2.266 a.u.) the computed potential curve goes over a maximum at 2.798 a.u., then descends rapidly. At R_e, 2σ is a B–H bonding MO, 3σ is a predominantly boron s–pσ hybrid AO, but including some $1s_H$, and 4σ is a 3s Rydberg AO localized on the boron. As R increases, a complicated metamorphosis occurs, in which 2σ changes to a $2s_B$ AO, 3σ finally becomes a $1s_H$ AO, and 4σ becomes a $2p\sigma_B$ AO. At intermediate distances (near 3.2 a.u.) 4σ is a very strongly BH-antibonding MO. A population analysis [48] gives highly anomalous populations in the 4σ MO: at the worst point (2.8 a.u.) the computed gross atomic population in 4σ is 1.28 in $1s_H$ and -0.38 in $2s_B$ (and $+0.10$ in $2p\sigma_B$). This is the worst-known example of the difficulties discussed in Section II.E, and is evidently associated with the existence of unusually strong negative (antibonding) overlap between the s_B and H STFs.

Pearson *et al.* [49] also have discussed the $1\sigma^2 2\sigma^2 3\sigma 4\sigma$, $^3\Sigma^+$ and $^1\Sigma^+$ states, using a smaller basis set but with electron correlation. As in the SCF calculation [48], they find a maximum followed by a repulsion curve in the $^3\Sigma^+$ state. The $^1\Sigma^+$ state also has a maximum, followed by a shallow second minimum before dissociation to an excited boron plus H.[13]

3. CH^+ and CH

On the $1\sigma^2 2\sigma^2 3\sigma^2$, $^1\Sigma^+$ ground state, the $1\sigma^2 2\sigma^2 3\sigma 4\sigma$, $^3\Sigma^+$ repulsive state, and the $1\sigma^2 2\sigma^2 3\sigma 1\pi$, $^3\Pi$ and A $^1\Pi$ states of CH^+, Green *et al.* [53] have done extensive CM calculations, and constructed derived potential curves for the states treated. They used SCF calculations with a large STF basis set, supplemented by configurations constructed by populating the virtual MOs obtained from the SCF calculations. However (before the CM), a two-configuration MCSCF calculation (configurations $1\sigma^2 2\sigma^2 3\sigma^2$ and $1\sigma^2 2\sigma^2 4\sigma^2$) was used for the ground state but ordinary SCF for the others. The numbers of configurations used for each state were 3370 ($^1\Sigma^+$), 3251 ($^3\Pi$), 3126 ($^1\Pi$), and 2890 ($^3\Sigma^+$). Green *et al.* made a few trials using NOs instead of SCF virtual MOs in setting up their excited configurations, but obtained slightly *worse* results. They conclude that the iterative natural orbital method of Bender and Davidson is not effective for improving wave functions of the quality they obtained.

From the wave functions of the $^1\Sigma^+$ and $^1\Pi$ states, Yoshimine *et al.* [54] have computed transition moments (cf. Section II.D) as a function of R, and band oscillator strengths for the $^1\Pi$–$^1\Sigma^+$ transition. In good agreement with their results are direct computations of transition moments and oscillator strengths by Martin *et al.* using the "equations of motion" method [55].

Some features of the $^3\Sigma^+$ curve are of interest. The curve is essentially repulsive [53] and does not show the occurrence of a Rydberg minimum, unlike the isoelectronic state of BH (see Section E.2). However, the curve shows a minimum at large R (of about 0.0013 a.u. depth at about 6.5 a.u.) attributable to induced-dipole attraction, proportional to R^{-4}, between C^+ and H. (All the states mentioned dissociate to $C^+ + H$.)

A very thorough study of the $1\sigma^2 2\sigma^2 3\sigma^2 1\pi$, $^2\Pi$ and the $1\sigma^2 2\sigma^2 3\sigma 1\pi^2$, $^4\Sigma^-$, $^2\Delta$, $^2\Sigma^-$, and $^2\Sigma^+$ states of CH (all experimentally known except $^4\Sigma^-$) has been made by Lie *et al.* [56]. With a basis set of 14σ, 8π, 4δ, and 2ϕ carbon and 9σ, 5π, and 2δ hydrogen STFs, they carried out MCSCF calculations followed by extended CM calculations for the states mentioned, and then constructed potential curves and calculated various other properties of these states. For the MCSCF calculations, they used for the $^2\Pi$ state four

[13] Here see *ab initio* calculations of Browne and Greenawalt [50] and of Houlden and Csizmadia [51]. Further, calculations on eight states of BH for vertical excitation have been made using many-body perturbation theory by Stern and Kaldor [52].

configurations necessary at large R to get correct dissociation, supplemented by four more which serve to provide partial correlation ($1s^2 2p^4$, ^3P in addition to $1s^2 2s^2 2p^2$, ^3P) in the C atom at $R = \infty$.

One configuration is enough to give correct dissociation in the cases of the $^4\Sigma^-$, $^2\Delta$, and $^2\Sigma^-$ states, but two are needed for the $^2\Sigma^+$; equal numbers of configurations were, however, added enlarging the MCSCFs to provide for the C atom CM at ∞.

For satisfactory results agreeing well with experiment it was found necessary to use "extended" CM, based on a consideration of configuration functions as follows [56]:

(a) those already described;
(b) all other CSFs arising from distributing five electrons among the valence MOs;
(c) all CSFs arising from distributing four electrons in valence orbitals and one electron in *external* MOs (MCSCF virtual MOs or NOs derived from these and the MCSCF MOs);
(d) all CSFs arising from distributing three electrons in valence MOs and two in external MOs, with certain exceptions.

Not included, however, were CSFs corresponding to the $1\sigma^2$ shell; it was argued that these would be unimportant for potential curves and other properties of especial interest, but see the comments in the first paragraph of Section E.1. A CM including all CSFs of types (a)–(c) has been called a "first-order" CM [57].

The calculations consisted of five steps:

(1) calculation of integrals;
(2) SCF or MCSCF calculation;
(3) determination of an approximate extended CM wave function using CSFs constructed from the full set of MCSCF occupied and virtual MOs, but approximating by zero all off-diagonal Hamiltonian matrix elements that involve only (c) and (d) CSFs;
(4) the NOs extracted from this wave function, ordered by symmetry and decreasing occupation numbers, then truncated to 13σ, 10π, 6δ, and 2ϕ NOs;
(5) the CM calculations.

Besides the full extended CM calculations, first-order CM and "valence" CM (here only the core and valence MOs were included) calculations were made, but only the extended CM calculations gave results in very good agreement with the experimental evidence. Only the extended CM calculations reproduced correctly the fact that the $^2\Sigma^-$ is a (weakly) bound state with a small maximum in its potential curve. Somewhat surprisingly, the results of the relatively simple valence CM calculation agreed better with experiment than those of the first-order CM approximation.

From the wave functions obtained, Lie, Hinze, and Liu give computed values of R_e, D_e, dipole moment, Hellmann–Feynman force, gradient of electric field at each nucleus, quadrupole moment, and infrared transition dipole moments and line strengths, with other details of interest.

4. OH

Although the most accurate calculation on the ground state of OH, but only for a somewhat limited range of R values, is that of Meyer [39] (see last part of Section D), some recent papers using the MCSCF method plus limited CM on the $1\sigma^2 2\sigma^2 3\sigma^2 1\pi^3$, X $^2\Pi$ and the $1\sigma^2 2\sigma^2 3\sigma 1\pi^4$, A $^2\Sigma^+$ low excited state deserve consideration since they extend over a range of R values out to dissociation.

One, a 14-term OVC calculation, by Stevens *et al.* [58] discusses OVC and MCSCF methods in some detail (see Section D) and their application to the ground state of OH. They obtain a good potential curve and a good curve of the dipole moment as a function of R. Their theoretical dissociation energy of 4.53 eV comes close to the experimental value of 4.63 eV. The computed dipole moment at R_e (1.674 D) agrees with the experimental value of 1.66 ± 0.01 D. Another MCSCF calculation, by Chu *et al.* [59], obtains similar but somewhat less accurate results on the X and A states, and includes also calculations of quadrupole moments and electric field gradients. More recently, Arnold *et al.* [60] have improved the calculation of Stevens *et al.* by starting with a 17-configuration OVC calculation, then adding CM to a total of 61 configurations. Their calculated D_e is 4.62 eV and the dipole moment at R_e is 1.637 D. In still another paper on the X and A states of OH, Meyer [61] uses the CEPA (coupled electron pair) approach to obtain extremely good agreement with experiment for the spectroscopic constants, which reflect the potential curve regions fairly near R_e. His results for R values out to dissociation, including dipole moment curves, are less accurate than the OVC results.

In general, even valence-shell MCSCF MOs can differ very appreciably from like-named SCF MOs. The differences for X $^2\Pi$ of OH are well illustrated by contour plots of the MOs (*not* of their charge densities) in the two cases. Such a comparison is shown in Fig. 7, from the paper of Stevens *et al.* [58]. Notably, the 2σ MOs change from bonding (SCF) to antibonding (MCSCF); here see also Table III of Ref. 58; and the 3σ MOs become more bonding (cf. discussion in Section C). Of interest also are the forms of the SCF-unoccupied MOs used in CM, some of which are shown in Fig. 8, from the same paper.

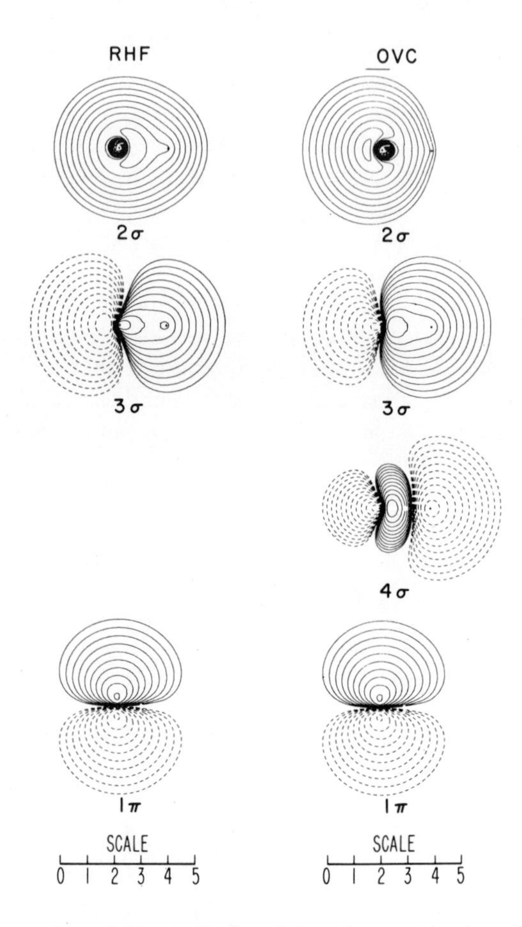

FIG. 7 Contour plots of the amplitudes of the valence molecular orbitals of the OVC III wave function at $R(\mathrm{OH}) = 1.8342$ bohrs. [From W. J. Stevens, G. Das, A. C. Wahl, M. Krauss, and D. Neumann, *J. Chem. Phys.* **61**, 3686 (1974).]

5. NH

Although the most accurate CM computation on the ground state of NH at R_e only is that of Bender and Davidson [26], 10-configuration OVC calculations by Stevens *et al.* [62] cover a range of values from 1.0 to 3.5 a.u., also 10 a.u., and yield a potential curve, spectroscopic constants, and the dipole moment as a function of R, also a dissociation energy $D = 3.37$ eV. The latter compares well with a rather uncertain experimental value of 3.40 eV. Calculations on the X $^3\Sigma^-$, a $^1\Delta$, b $^1\Sigma^+$, A $^3\Pi$, and c $^1\Pi$ states of NH by the POL–CI and other methods are reported by Hay and Dunning [63].

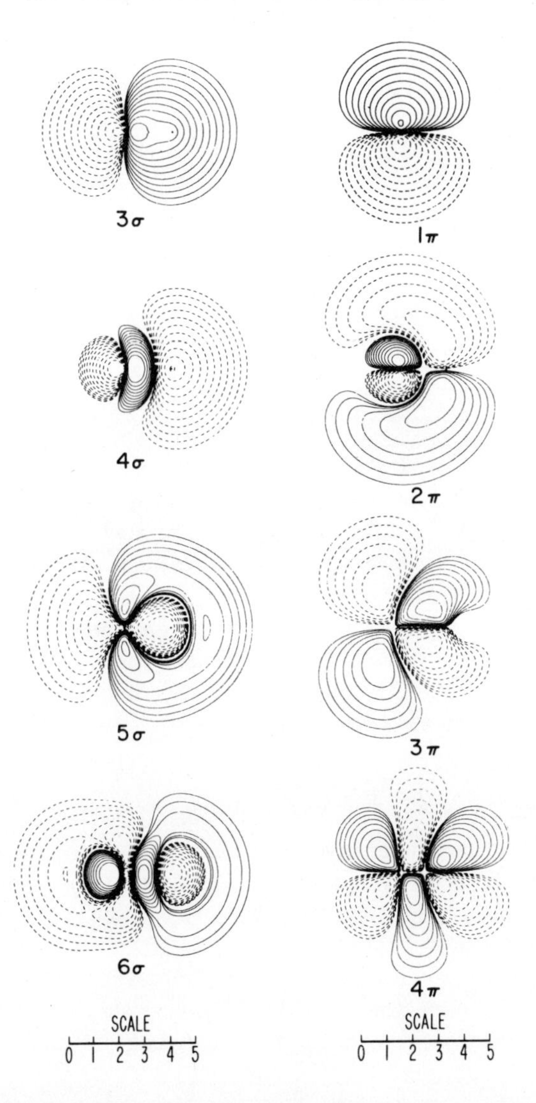

FIG. 8. Contour plots of the amplitudes of the valence orbitals of the OVC XIV wave function at $R(\mathrm{OH}) = 1.8342$ bohrs. [From W. J. Stevens, G. Das, A. C. Wahl, M. Krauss, and D. Neumann, *J. Chem. Phys.* **61**, 3686 (1974).]

6. VH and HgH

Henderson *et al.* [64] have made OVC or MCSCF plus CM calculations on a large number of bound states of VH. Potential curves and spectroscopic parameters are obtained. The lowest group of bound states, all dissociating to normal atoms V (d^3s^2, ^4F)+H, are $^5\Delta$ (lowest), $^5\Pi$, $^5\Sigma^-$, $^5\Phi$, $^3\Delta$, $^3\Phi$, $^3\Sigma^-$, and $^3\Pi$. Above these, potential curves for eleven other bound states dissociating to excited V plus H are shown. The calculated dissociation energy for the lowest state is about 1.8 eV.

For the ground ($^2\Sigma^+$) and low excited ($^2\Pi$) states of HgH, Das and Wahl [65] use a modified pseudopotential and MCSCF approach to obtain potential curves.

7. NaH

Sachs *et al.* [66] have made MCSCF calculations and other properties of six states of NaH over a wide range of internuclear distances. However, only the two valence electrons were correlated. A rotation–vibrational analysis was made for those states (X $^1\Sigma^+$, A $^1\Sigma^+$, and b $^3\Sigma^+$) which are bound, also for NaD. Transition moments and line and band strengths for spectroscopic transitions are also calculated. Wing broadening of the sodium D lines in Na+H collisions was also studied. Finally, a pseudopotential method was tested.

8. GVB Calculations

Goddard *et al.* have made several calculations on diatomic hydrides using methods of the GVB type (see the end of Section III.D). Two papers discuss ground and excited states of LiH [67]. BH is discussed in another paper [68]. Hay and Dunning [63] have used the polarization configuration mixing (POL–CI) method, starting from GVB, in a discussion of several low energy states of NH. Dunning has discussed low-lying states of HF in a similar manner [69]. All these papers have the advantage of carrying the calculations out to dissociation.

REFERENCES

1. P. E. Cade and W. H. Huo, *J. Chem. Phys.* **47**, 614–649 (1967); *Atomic Data* **12**, 415 (1973).
2. W. G. Richards, T. E. H. Walker, and R. K. Hinkley, "A Bibliography of *Ab Initio* Molecular Wave Functions," Oxford Univ. (Clarendon) Press, London and New York, 1971; W. G. Richards, T. E. H. Walker, L. Farnell, and P. R. Scott, "Supplement for 1970–1973," Oxford Univ. (Clarendon) Press, London and New York, 1974.

3. (a) V. Bondybey, P. K. Pearson, and H. F. Schaefer, III, *J. Chem. Phys.* **57**, 1123 (1973).
(b) P. Sutton, B. Bertoncini, G. Das, T. L. Gilbert, and A. C. Wahl, *Int. J. Quantum Chem.* **3S**, 1344 (1973).
4. P. S. Bagus and H. F. Schaefer, III, *J. Chem. Phys.* **58**, 1844 (1973).
5. R. S. Mulliken, *Int. J. Quantum Chem.* **5**, 95 (1971).
6. R. S. Mulliken, *Acc. Chem. Res.* **9**, 7 (1976).
7. R. S. Mulliken, *Chem. Phys. Lett.* **14**, 141 (1972).
8. W. H. Henneker and H. Popkie, *J. Chem. Phys.* **54**, 1763 (1971).
9. (a) TiH and VH: P. R. Scott and W. G. Richards, *J. Phys. B* **7**, 500, 1347 (1974); ScH: *ibid.* **7**, 1679 (1974).
(b) P. S. Julienne, M. Krauss, and A. C. Wahl, *Chem. Phys. Lett.* **11**, 16 (1971).
10. (a) R. F. W. Bader, I. Keaveney, and P. E. Cade, *J. Chem. Phys.* **47**, 3381 (1967).
(b) P. E. Cade, R. F. W. Bader, W. H. Henneker, and I. Keaveney *J. Chem. Phys.* **50**, 5313 (1969).
(c) R. F. W. Bader, *in* "International Review of Science: Theoretical Chemistry: Physical Chemistry" (A. D. Buckingham and C. A. Coulson, eds.), Ser. 2, Vol. 1, pp. 43–78. Butterworth, London, 1975.
11. P. E. Cade, R. F. W. Bader, and J. Pelletier, *J. Chem. Phys.* **54**, 3517 (1971).
12. R. F. W. Bader and W. H. Henneker, *J. Am. Chem. Soc.* **88**, 280 (1966).
13. (a) R. F. W. Bader and P. M. Beddall, *J. Chem. Phys.* **56**, 3320 (1972); **58**, 557 (1973); S. Srebenik and R. W. Bader, *ibid.* **61**, 2536 (1974).
(b) R. F. W. Bader and P. M. Beddall, *J. Am. Chem. Soc.* **95**, 305 (1973); R. F. W. Bader and R. R. Messer, *Can. J. Phys.* **52**, 2268 (1974).
14. A. M. Karo, *J. Chem. Phys.* **31**, 182 (1959).
15. R. S. Mulliken, *J. Chem. Phys.* **23**, 1841, 2338, 2343 (1955).
16. R. S. Mulliken, *in* "Quantum Theory of Atoms, Molecules, and the Solid State" (A Tribute to J. C. Slater) (P.-O. Löwdin, ed.), pp. 231–241. Academic Press, New York, 1966.
17. C. W. Kern and M. Karplus, *J. Chem. Phys.* **40**, 1374 (1964).
18. R. K. Nesbet, *J. Chem. Phys.* **36**, 1518 (1962).
19. E. Clementi, *J. Chem. Phys.* **36**, 33 (1962).
20. P. Politzer and R. S. Mulliken, *J. Chem. Phys.* **55**, 5135 (1971), Table II.
21. (a) K. R. Roby, *Mol. Phys.* **27**, 81 (1974).
(b) R. Heinzmann and R. Ahlrichs, *Theor. Chim. Acta* **42**, 33 (1976).
22. R. S. Mulliken, *J. Chem. Phys.* **36**, 3428 (1962).
23. P. Politzer and R. R. Harris, *J. Am. Chem. Soc.* **92**, 6451 (1968).
24. A. M. Karo, *J. Chem. Phys.* **31**, 182 (1959).
25. S. Fraga and B. J. Ransil, *J. Chem. Phys.* **36**, 1127 (1962).
26. C. F. Bender and E. R. Davidson, *Phys. Rev.* **183**, 23 (1969).
27. C. F. Bender and E. R. Davidson, *J. Phys. Chem.* **70**, 2675 (1966).
28. W. J. Stevens, *J. Chem. Phys.* **58**, 1264 (1973); M. Krauss and D. Neumann, *Mol. Phys.* **27**, 917 (1974).
29. (a) A. D. McLean and B. Liu, *J. Chem. Phys.* **38**, 1066 (1973).
(b) P. S. Bagus, B. Liu, A. D. McLean, and M. Yoshimine, *in* "Wave Mechanics: The First Fifty Years" (W. C. Price, S. S. Chissich, and T. Ravensdale, eds.), Chapter 8. Butterworth, London, 1973.
30. W. J. Stevens, G. Das, A. C. Wahl, M. Krauss, and D. Neumann, *J. Chem. Phys.* **61**, 3686 (1974); F. P. Billingsley, II, and M. Krauss, *ibid.* **60**, 4130 (1974).
31. M. Cohen and A. Dalgarno, *Proc. Phys. Soc. London* **77**, 748 (1961); G. G. Hall, *Phil. Mag.* **6**, 249 (1961).

32. P. E. Cade and W. M. Huo, *J. Chem. Phys.* **45**, 1063 (1966).
33. (a) S. Green, *J. Chem. Phys.* **54**, 827 (1971).
    (b) S. Green, *Adv. Chem. Phys.* **25**, 179 (1974).
34. (a) C. Moeller and M. S. Plesset, *Phys. Rev.* **46**, 618 (1934).
    (b) C. W. Kern and M. Karplus, *J. Chem. Phys.* **40**, 1374 (1964).
35. K. E. Banyard and M. R. Hayns, *J. Phys. Chem.* **75**, 416 (1971).
36. (a) G. C. Lie and E. Clementi, *J. Chem. Phys.* **60**, 1275 (1974).
    (b) G. C. Lie and E. Clementi, *J. Chem. Phys.* **60**, 1288 (1974).
37. W. Meyer, *Int. J. Quantum Chem.* **55**, 341 (1971).
38. W. Meyer, *J. Chem. Phys.* **58**, 1017 (1973).
39. W. Meyer, *Theor. Chim. Acta.* **35**, 277 (1974).
40. W. Meyer and P. Rosmus, *J. Chem. Phys.* **63**, 2356–2375 (1975).
41. W. Meyer, *J. Chem. Phys.* **64**, 290 (1976); C. E. Dysktra, H. F. Schaefer, III, and W. Meyer, *ibid.* **65**, 2740 (1976).
42. S. F. Boys and N. C. Handy, *Proc. Roy. Soc. London* **311A**, 309 (1969).
43. C. F. Bender and E. R. Davidson, *J. Chem. Phys.* **49**, 4222 (1968).
44. K. K. Docken and J. Hinze, *J. Chem. Phys.* **57**, 4928, 4936 (1972).
45. R. F. Stewart, D. K. Watson, and A. Dalgarno, *J. Chem. Phys.* **63**, 3222 (1975).
46. P. S. Bagus, C. M. Moser, P. Goethals, and G. Verhaegen, *J. Chem. Phys.* **58**, 1886 (1973).
47. F. Grimaldi, A. Lecourt, H. Lefebvre-Brion, and C. M. Moser, *J. Mol. Spectros.* **20**, 341 (1966).
48. R. S. Mulliken, *Int. J. Quantum Chem.* **5**, 83 (1971).
49. P. K. Pearson, C. F. Bender, and H. F. Schaefer, III, *J. Chem. Phys.* **55**, 5235 (1971).
50. J. C. Browne and E. M. Greenawalt, *Chem. Phys. Lett.* **7**, 363 (1970).
51. S. A. Houlden and G. Csizmadia, *Theor. Chim. Acta.* **35**, 173 (1974).
52. P. S. Stern and U. Kaldor, *J. Chem. Phys.* **64**, 2002 (1976).
53. S. Green, P. S. Bagus, B. Liu, A. D. McLean, and M. Yoshimine, *Phys. Rev. A* **5**, 1614 (1972).
54. M. Yoshimine, S. Green, and P. Thaddeus, *Astrophys. J.* **183**, 899 (1973).
55. P. H. S. Martin, D. L. Yeager, and V. McKoy, *Chem. Phys. Lett.* **25**, 182 (1974).
56. G. C. Lie, J. Hinze, and B. Liu, *J. Chem. Phys.* **59**, 1872, 1887 (1973); **57**, 625 (1972).
57. H. F. Schaefer, III, R. A. Klemm, and F. E. Harris, *Phys. Rev.* **181**, 137 (1969).
58. W. J. Stevens, G. Das, A. C. Wahl, M. Krauss and D. Nuemann, *J. Chem. Phys.* **61**, 3686 (1974).
59. S. I. Chu, M. Yoshimine, and B. Liu, *J. Chem. Phys.* **61**, 5389 (1974).
60. J. O. Arnold, E. E. Whiting, and L. F. Sharbaugh, *J. Chem. Phys.* **64**, 3251 (1976).
61. W. Meyer, *Theor. Chim. Acta.* **35**, 277 (1974).
62. W. J. Stevens, *J. Chem. Phys.* **58**, 1264 (1973); G. Das, A. C: Wahl, and W. J. Stevens, *ibid.* **61**, 433 (1974).
63. P. J. Hay and T. H. Dunning, Jr., *J. Chem. Phys.* **64**, 5077 (1976).
64. G. A. Henderson, G. Das, and A. C. Wahl, *J. Chem. Phys.* **63**, 2805 (1975).
65. G. Das and A. C. Wahl, *J. Chem. Phys.* **64**, 4672 (1976).
66. E. S. Sachs, J. Hinze, and N. H. Sabelli, *J. Chem. Phys.* **62**, 3367, 3377, 3384, 3389, 3393 (1975).
67. W. E. Palke and W. A. Goddard, III, *J. Chem. Phys.* **50**, 4524 (1969); C. F. Melius and W. A. Goddard, III, *ibid.* **56**, 3348 (1973).
68. R. J. Blint and W. A. Goddard, III, *J. Chem. Phys.* **57**, 5296 (1972); *Chem. Phys. Lett.* **14**, 616 (1972).
69. T. H. Dunning, Jr., *J. Chem. Phys.* **65**, 3854 (1976).

CHAPTER V

HOMOPOLAR DIATOMIC MOLECULES

A. SCF CALCULATIONS

Other than on H_2, the first SCF calculation on a homopolar diatomic molecule was that by Scherr [1a] on N_2 in Roothaan's laboratory at Chicago, using Roothaan's LCAO–SCF method. The difficult interelectronic repulsion integrals were evaluated by methods devised principally by Roothaan and Ruedenberg. Scherr used a minimal basis set, with Slater's atomic ζ's. The work, done by Scherr with the help of two assistants, using hand computers, took two years. Scherr dealt not only with the ground state $1\sigma_g{}^2 1\sigma_u{}^2 2\sigma_g{}^2 2\sigma_u{}^2 1\pi_u{}^4 3\sigma_g{}^2$, ${}^1\Sigma_g{}^+$ but also with the (vertical) excitation energies of the ${}^3\Pi_g$ and ${}^1\Pi_g$ excited states of the configuration $\ldots 1\pi_u{}^4 3\sigma_g 1\pi_g$, the ${}^3\Pi_u$ state of the configuration $\ldots 2\sigma_u 1\pi_u{}^4 3\sigma_g{}^2 1\pi_g$, and the six excited states ${}^3\Sigma_u{}^+$, ${}^3\Delta_u$, ${}^3\Sigma_u{}^-$, ${}^1\Sigma_u{}^-$, ${}^1\Delta_u$, and ${}^1\Sigma_u{}^+$ of the configuration $\ldots 1\pi_u{}^3 3\sigma_g{}^2 1\pi_g$. His results for the ground state are shown in Table 1. Scherr also computed the quadrupole moment of N_2 from his functions, and carried out a population analysis. Scherr's paper contains instructive tables giving the values of the various attraction and repulsion integrals which go into his SCF calculations.

In his calculations on the excited states Scherr used a method suggested

TABLE 1

Basis Set, ε Values, and Coefficients for MOs of Ground State of N_2 at R_e (2.068 a.u.)[a–c]

Basis	MO: $-\varepsilon$:	$1\sigma_g$ 15.722	$1\sigma_u$[d] 15.720	$2\sigma_g$ 1.453	$2\sigma_u$ 0.731	$1\pi_u$ 0.580	$3\sigma_g$ 0.545
σ_g 1s		0.999		−0.082			−0.043
σ_g 2s		0.012		0.672			−0.589
σ_g 2p		0.003		0.339			0.853
σ_u 1s			1.002		0.011		
σ_u 2s			0.027		1.020		
σ_u 2p			0.012		−0.376		
π_u 2p						0.883	

[a] Computed total energy $E = -108.574$ a.u. from Scherr [1a].

[b] The basis functions are simple partially normalized LCSTFs of the form $2^{-1/2}(\chi_a^{\zeta} \pm \chi_b^{\zeta})$.

[c] The ζ values for the STFs were those proposed by Slater for the N atom [J. C. Slater, *Phys. Rev.* **36**, 57 (1930); C. Zener, *ibid.* **36**, 51 (1931)]: $\zeta = 6.7$ for 1s and 1.95 for 2s, $2p\sigma$, and $2p\pi$.

[d] Scherr gives 17.720 for $\varepsilon(1\sigma_u)$, but we assume that this was a misprint.

by Roothaan, who pointed out that the vertical excitation energy for excitation from a nondegenerate occupied MO to an occupied MO for a closed-shell molecule is given approximately by

$$\Delta E_{\text{exc}} = \varepsilon_a - \varepsilon_i + (J_{ia} - K_{ia}) \pm K_{ia}. \tag{1}$$

Here ε_i is the SCF orbital energy of the initially occupied MO, and ε_a is the virtual SCF orbital energy (see Section I.C) of the excited MO, and J_{ia} and K_{ia} are the Coulomb and exchange integrals between MOs ϕ_i and φ_a. The excitation energies of the Π states mentioned previously were computed by Eq. (1); in this equation the $\pm$ sign refers to the singlet and triplet states respectively. For the Σ and Δ excited states already mentioned, $\varepsilon_a - \varepsilon_i$ must be supplemented by more elaborate expressions than in Eq. (1); these expressions are given in Scherr's paper, and also in an earlier paper by Recknagel.[1]

Others in Roothaan's laboratory, notably B. J. Ransil, P. E. Cade, and A. C. Wahl, carried on the work on diatomic molecules with steadily increasing sophistication. In a recent paper in *Atomic Data* [2], Cade and Wahl have tabulated the results of the best available SCF calculations on first-row homopolar diatomic molecules and on some of their ions and excited states. They give results on H_2, Li_2, Li_2^+ (two states), Be_2, Be_2^+

[1] For the excitation energies of the states of configuration $\ldots 1\pi_u^3 3\sigma_g 1\pi_g$, see Recknagel [1b].

(two states), B_2 (three states), B_2^+, B_2^-, C_2 (five states), C_2^+, C_2^- (three states), N_2, N_2^+ (three states), N_2^{2+}, O_2 (three states), O_2^+, O_2^-, F_2, F_2^+ (four states), and F_2^{2+}. Table 2 summarizes the ground-state results on N_2; here comparison with Table 1 is of interest. Note the improvement of 0.419 a.u. in the computed total energy E. For comparison with Table 2 on N_2, Tables 4 and 5 are taken from Cade and Wahl's tables on Li_2 and O_2.

TABLE 2

Basis Set for the Ground State of N_2 at its R_e (2.068 a.u.), Followed by ε Values and STF Coefficients for the Occupied MOs[a–c]

LCSTF	MO: $-\varepsilon$:	$1\sigma_g$ 15.682	$1\sigma_u$ 15.678	$2\sigma_g$ 1.474	$2\sigma_u$ 0.778	$1\pi_u$ 0.615	$3\sigma_g$ 0.635
σ_g 1s (5.683)[d]		0.923		−0.279			0.075
σ_g 1s (10.342)		0.152		−0.006			0.003
σ_g 2s (1.453)		0.001		0.141			−0.459
σ_g 2s (2.439)		−0.000		0.599			−0.177
σ_g 3s (7.040)		−0.085		−0.023			−0.007
σ_g 2p (1.283)		0.000		0.116			0.429
σ_g 2p (2.570)		0.001		0.259			0.485
σ_g 2p (6.217)		0.001		0.011			0.025
σ_g 3d (1.341)		0.000		0.036			0.047
σ_g 3d (2.917)		0.001		0.040			0.031
σ_g 3d (5.521)		−0.000		−0.003			−0.002
σ_g 4f (2.594)		0.000		0.013			0.011
σ_u 1s (5.955)			0.934		−0.244		
σ_u 1s (10.659)			0.155		−0.000		
σ_u 2s (1.570)			−0.012		0.364		
σ_u 2s (2.490)			0.005		0.547		
σ_u 3s (7.292)			−0.053		−0.031		
σ_u 2p (1.485)			−0.007		−0.414		
σ_u 2p (3.500)			0.003		−0.109		
σ_u 3d (1.690)			−0.001		−0.036		
π_u 2p (1.384)						0.469	
π_u 2p (2.533)						0.399	
π_u 2p (5.692)						0.031	
π_u 3d (2.057)						0.059	
π_u 3d (2.707)						0.017	
π_u 4f (3.069)						0.012	

[a] Computed total energy $E = -108.9928$ a.u.

[b] See footnote *b* to Table 1.

[c] The ζ and ε values and coefficients, given by Cade to five figures after the decimal point, are here rounded to three figures.

[d] ζ Values of the STFs in parentheses.

Table 3 contains the results of a new calculation for N_2 using an even-tempered basis set (see Section II.A). We used the augmented-triple-ζ atomic basis set of Raffenetti [3] plus three 3d-type and one 4f-type STF as polarization functions. The parameters used to define the atomic STFs are $\alpha_{1s} = 0.822525$, $\beta_{1s} = 1.673719$, $\alpha_{2p} = 0.731294$, and $\beta_{2p} = 1.813350$ [3]. The polarization functions, optimized for the molecule, are $\alpha_{3d} = 0.65$, $\beta_{3d} = 2.00$, $\alpha_{4f} = 2.75$, and $\beta_{4f} = 1.00$. The wave functions could be further improved through reoptimization of all the α's and β's. We have included the explicit values of the exponential factors in Table 3, although they are obtainable from the (α_l, β_l). Note that the same exponents were used for the g as for the u 1s STFs and for the g and u, σ and π, STFs of each of the types 2p, 3d, and 4f. The use of even-tempered basis sets for molecular calculations shows promise in that fewer parameters need to be optimized (two for each l type), and only the lowest n value of a given l type is required (as is

TABLE 3

Even-Tempered Basis Set of STFs for the Ground State of N_2 ($R = 2.068$ a.u.), Followed by ε Values and STF Coefficients for the Occupied MOs[a, b]

STF	MO:	$1\sigma_g$	$1\sigma_u$	$2\sigma_g$	$2\sigma_u$	$1\pi_u$	$3\sigma_g$
	$-\varepsilon$:	15.679	15.676	1.472	0.777	0.614	0.634
1s (1.377)[c]		0.001	0.000	0.349	0.849		0.753
1s (2.504)		−0.001	0.000	0.483	0.037		−0.417
1s (3.857)		0.020	−0.018	−0.558	−0.346		0.097
1s (6.455)		0.647	−0.648	−0.014	−0.074		−0.110
1s (10.803)		0.046	−0.046	−0.021	−0.002		0.022
1s (18.082)		0.000	0.000	0.002	−0.001		−0.004
2p (1.326)		0.001	0.000	0.068	−0.147	0.294	−0.316
2p (2.405)		−0.001	0.001	0.180	−0.146	0.305	−0.297
2p (4.361)		0.001	−0.001	0.019	−0.021	0.030	−0.055
2p (7.907)		0.000	0.000	0.002	−0.002	0.005	−0.001
3d (1.300)		0.000	0.000	0.019	−0.005	0.022	−0.041
3d (2.600)		0.001	0.000	0.032	−0.006	0.041	−0.023
3d (5.200)		0.000	0.000	0.000	0.000	−0.003	0.000
4f (2.750)		0.000	0.000	0.010	0.001	0.010	−0.010

[a] Computed $E = -108.9913$ a.u.

[b] The basis functions labeled 1s, 2p, 3d, and 4f refer to symmetrized combinations such as σ_g1s and σ_u1s $(\chi_a \pm \chi_b)$ of STFs χ_a and χ_b; but the coefficients are those of the individual STFs (either χ_a or χ_b). For example, columns 2, 3, and 6 of the last row contain coefficients for the basis functions $2^{-1/2}(4f\sigma_a - 4f\sigma_b)$, $2^{-1/2}(4f\sigma_a + 4f\sigma_b)$, and $2^{-1/2}(4f\pi_a + 4f\pi_b)$ respectively.

[c] ζ Values in parentheses.

TABLE 4

Basis Set for Ground State of Li_2 at its R_e (5.051 a.u.), followed by ε values and STF Coefficients[a, b]

LCSTF	MO: $-\varepsilon$:	$1\sigma_g$ 2.452	$1\sigma_u$ 2.452	$2\sigma_g$ 0.182
σ_g 1s (2.264)[c]		0.897		−0.191
σ_g 1s (4.471)		0.166		−0.015
σ_g 2s (0.665)		−0.000		0.773
σ_g 3s (1.533)		0.001		0.002
σ_g 3s (2.703)		−0.060		−0.034
σ_g 3s (3.977)		−0.002		0.004
σ_g 2p (0.740)		0.000		0.157
σ_g 2p (1.500)		0.002		−0.001
σ_g 2p (2.287)		−0.009		0.004
σ_g 2p (2.971)		0.003		0.002
σ_g 3d (1.151)		0.000		0.009
σ_u 1s (2.438)			0.755	
σ_u 1s (4.336)			0.195	
σ_u 2s (1.026)			0.002	
σ_u 2s (3.049)			0.074	
σ_u 3s (2.006)			0.002	
σ_u 2p (0.852)			0.001	
σ_u 2p (1.909)			−0.005	

[a] Computed $E = -14.8715$ a.u.
[b] See footnote b to Table 1.
[c] ζ Values in parentheses.

also true for GTFs). In the present case six 1s, four 2p, three 3d, and one f STF (44 in total) yielded an energy only 0.0015 a.u. higher than in Table 2 which used a basis of twelve σ_g, eight σ_u, and six π_u functions (52 in total), a majority of which were individually optimized.

Among other SCF calculations, reference should be made to a paper by Gilbert and Wahl on wave functions and potential curves foı low-lying states of He_2^+, Ne_2^+, Ar_2^+, F_2, F_2^-, and the ground state of Cl_2 [4]. From SCF calculations Henderson, Zemke, and Wahl have obtained potential curves, spectroscopic constants, and one-electron properties for the lowest $^2\Sigma_g^+$ and $^2\Pi_u$ states of Li_2^+ [5]. Müller and Jungen have calculated potential curves for several excited states of Li_2^+ [6].

Of considerable interest is a so-called *correlation diagram* or binding scheme showing how the orbital energies ε of the MOs change in going from R_e (or larger R) to $R = 0$. (The word correlation here is being used in a very different sense than in "electron correlation.") Crude qualitative correlation diagrams for first-row diatomic molecules were presented long ago [7].

More recently, Mulliken has made approximate SCF calculations on N_2 resulting in a correlation diagram for the ground state between $R = 0$ and R_e [8]. The basis set used was double-ζ (two STFs for each one of a minimal set) plus one 3dσ and one 3dπ, except that some changes were made at the smallest R values.

TABLE 5

Basis Set for Ground State of O_2 ($1\sigma_g^2 1\sigma_u^2 2\sigma_g^2 2\sigma_u^2 3\sigma_g^2 1\pi_u^4 1\pi_g^2$, $^3\Sigma_g^-$) at Its R_e (2.282 a.u.), Followed by ε Values and STF Coefficients[a, b]

LCSTF	MO: $1\sigma_g$ $-\varepsilon$: 20.730	$1\sigma_u$ 20.729	$2\sigma_g$ 1.649	$2\sigma_u$ 1.099	$3\sigma_g$ 0.736	$1\pi_u$ 0.705	$1\pi_g$ 0.532
σ_g 1s (6.981)[c]	0.947		−0.261		0.079		
σ_g 1s (12.404)	0.093		0.003		−0.001		
σ_g 2s (1.614)	−0.001		0.205		−0.329		
σ_g 2s (2.762)	0.004		0.628		−0.227		
σ_g 3s (8.573)	−0.044		−0.028		0.000		
σ_g 2p (1.475)	−0.000		0.121		0.472		
σ_g 2p (2.889)	0.000		0.154		0.447		
σ_g 2p (5.917)	0.002		0.013		0.047		
σ_g 3d (1.710)	−0.000		0.055		0.071		
σ_g 3d (3.350)	0.000		0.013		0.016		
σ_g 4f (2.298)	−0.000		0.021		0.022		
σ_u 1s (7.319)		0.924		−0.233			
σ_u 1s (11.869)		0.080		−0.004			
σ_u 2s (1.992)		−0.006		0.790			
σ_u 3s (4.362)		0.008		0.284			
σ_u 2p (1.745)		−0.004		−0.215			
σ_u 2p (3.864)		0.003		0.091			
σ_u 3d (2.176)		−0.001		−0.024			
π_u 2p (1.353)						0.276	
π_u 2p (2.123)						0.464	
π_u 2p (3.769)						0.211	
π_u 2p (8.143)						0.011	
π_u 3d (2.221)						0.047	
π_u 3d (3.277)						0.006	
π_u 4f (2.878)						0.013	
π_g 2p (1.468)							0.634
π_g 2p (2.884)							0.512
π_g 2p (5.791)							0.058
π_g 3d (1.476)							−0.044
π_g 4f (2.385)							−0.012

[a] Computed $E = -149.6659$ a.u.
[b] See foot note b to Table 1.
[c] ζ Values in parentheses.

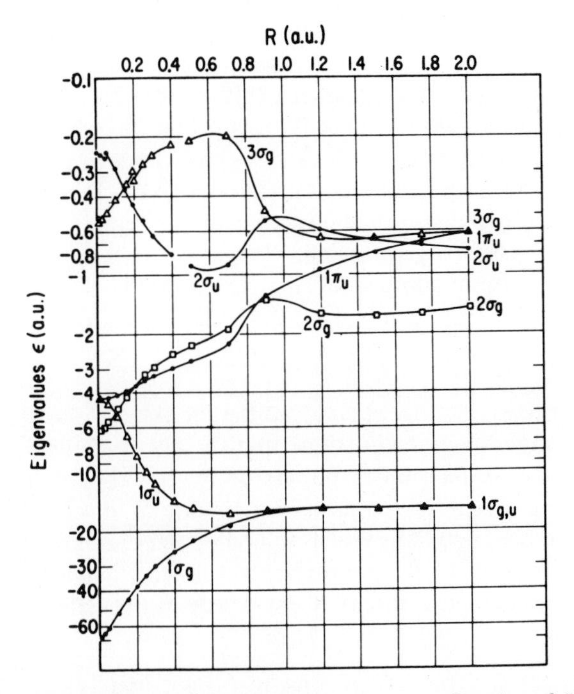

FIG. 1 Correlation diagram for N_2 (orbital energies ε as a function of R). [From R. S. Mulliken, *Int. J. Quantum Chem.* **8**, 817 (1974).]

The results are shown graphically in Fig. 1. A few of their features will be selected for emphasis: proceeding from R_e toward smaller R values,

(a) the ε values of $1\sigma_g$ and $1\sigma_u$ remain close together to about 1 a.u., after which $1\sigma_g$ decreases rapidly until at $R = 0$ it becomes 1s of the united-atom silicon, while $1\sigma_u$ increases somewhat to become $2p\sigma$ of the united atom;

(b) $2\sigma_g$ varies irregularly at first, but finally decreases strongly to become 2s of the united atom;

(c) $2\sigma_u$ at first varies irregularly, but finally increases to become $3p\sigma$ of the united atom;

(d) $1\pi_u$ decreases steadily, finally with increasing speed, to become $2p\pi$ of the united atom;

(e) $3\sigma_g$ mostly increases at first, from $-\varepsilon = 0.63$ at R_e, to a maximum value of $-\varepsilon = 0.19$ at about 0.6 a.u., where $3\sigma_g$ has now become almost pure 3s, and its $-\varepsilon$ is comparable with that (0.25) of 3s of magnesium, the semi-united atom (see Section C); remaining 3s all the way, it then decreases in energy until $R = 0$ (now $-\varepsilon = 0.55$).

It is seen that resemblance to the semi-united atom is closest near $R = 0.6$, while at $R = 0$ the united atom is reached. A population analysis of the MOs shows changing s and p character (plus a little d); in $1\sigma_g$ the p population reaches a maximum of about 0.15e at about 0.1 a.u., where in shape it resembles $1\sigma_g$ of H_2 at R_e of H_2, although of course it is much smaller.

Although LCSTF expressions with separate-atom STFs (SASTFs) can be used, with some qualifications, down to extremely small R values (such as 0.01 a.u.) [8], it is better for small R values to introduce UAOs (united-atom AOs) or corresponding UASTFs (each UAO is closely approximated by a UA LCSTF). Best, one can use *both* UASTFs and SASTFs [9, 10].[2] This has been done for He_2, N_2, and CO [10]. The results for CO resemble those for N_2.

In this connection, the viewpoint that the molecule at small R values is a perturbed atom can be used to show that the *electronic energy* can be approximated by an expansion in powers of R: $E = E_0 + E_2 R^2 + E_3 R^3 + \cdots$, where E_0 is the energy of the united-atom state at $R = 0$; this expression is applicable to heteropolar as well as to homopolar diatomic molecules [11]. Expansions in powers of R can also be used for the individual MO ε's.

The correlation diagram discussed previously can be extended to R values greater than R_e:

(a) to $R = \infty$ for *inner shells*, thus in N_2 $1\sigma_g$ and $1\sigma_u$ become $1s_a \pm 1s_b$ and so correlate with SA 1s, and $2\sigma_g$ and $2\sigma_u$ become $2s_a \pm 2s_b$ and correlate with SA 2s (by "inner shells" is meant all MO shells which correspond on dissociation to closed shells of the two separate atoms);

(b) for the *outer shells* $1\pi_u{}^4$ and $3\sigma_g{}^2$, correlation toward $2p\pi_a + 2p\pi_b$, or SA $2p\pi$, and to $2p\sigma_a - 2p\sigma_b$, or SA $2p\sigma$, makes sense to R values (say about $1.5R_e$) where the SCF approximation is still fairly good, but at larger R where extensive electron correlation becomes essential, these correlations fail to make sense, but one may speak of them as pseudocorrelations [12].

This distinction was not made in the early correlation diagrams [7].

B. ORBITAL ENERGIES AND IONIZATION ENERGIES

As was mentioned in Section III.C, $-\varepsilon$ values are expected to be rather good approximations to ionization energies I (Koopmans' theorem). Agreement should usually be somewhat better for so-called *vertical* I's, I_{vert}, than for *adiabatic* or ordinary I's, I_{ad}. Vertical I's correspond to an ionization process in which R is kept at R_e of the original molecule, while for adiabatic I's the transition is to a positive ion at its new R_e. For nonbonding

[2] Ermler *et al.* [10] include work on He_2, H_2, CO, and their positive ions.

MOs, the two differ little, whereas for bonding and antibonding MOs, R_e is respectively larger or smaller for the positive ion. But exact agreement of I_{vert}'s with $-\varepsilon$'s is not to be expected, because

(a) all the MOs are somewhat changed and the total energy is decreased (relaxation) in the process of ionization;

(b) the correlation (plus relativistic) energy is somewhat different in the ion than in the neutral molecule.

Discrepancy (a) can be removed in the SCF approximation if separate SCF calculations are made for molecule and ion, and the difference is taken. Discrepancy (b) can be dealt with if correlation energies can be computed or estimated; then values of I_{vert} and I_{ad} agreeing with experiment should be obtained.

Tables 6 and 7 compare $-\varepsilon$ values with experimental I's for the molecules

TABLE 6

Comparison of $-\varepsilon$ and I values for N_2 [a]

Orbital ionized	Ion state	I_{ad} (eV)	I_{vert} (eV)	$-\varepsilon$ (eV)
1s	$^2\Sigma^+$	409.5	409.9	426.71 ($1\sigma_g$)
				426.61 ($1\sigma_u$)
$2\sigma_g$	$^2\Sigma_g{}^+$		37.3	40.10
$2\sigma_u$	$^2\Sigma_u{}^+$	18.73	18.6	21.17
$1\pi_u$	$^2\Pi_u$	16.96	16.8	16.75
$3\sigma_g$	$^2\Sigma_g{}^+$	15.58	15.5	17.28

[a] I_{ad} are from experimental optical data, I_{vert} are experimental data from K. Siegbahn *et al.*, ESCA Applied To Free Molecules, North-Holland Publ., Amsterdam 1969. The ε values are from Cade and Wahl [2].

N_2 and O_2. For the most part, the magnitudes of the ε's are in the same order as the I's. An exception occurs in the case of the $3\sigma_g$ and $1\pi_u$ MOs of N_2. Moreover, the discrepancy is *increased* when the SCF-calculated energies of the ionic states $^2\Sigma_g{}^+$ and $^2\Pi_u$ obtained by respectively removing a $3\sigma_g$ or $1\pi_u$ electron are calculated: $^2\Pi_u$ is computed by SCF to lie appreciably below $^2\Sigma_g{}^+$, whereas experimentally the order is reversed. Thus cause (a) is more than inadequate to explain the discrepancy. This case, however, is a rather unusual one. It must be explained by the occurrence in cause (b) of a very considerably larger correlation energy for the $^2\Sigma_g{}^+$ than for the $^2\Pi_u$ ion.

TABLE 7

Comparison of $-\varepsilon$ and I Values for O_2 [a, b]

Orbital ionized	Ion state	I_{ad} (eV)	I_{vert} (eV)	$-\varepsilon$ (eV)	$\Delta E_{SCF}(D_{\infty h})$	$\Delta E_{SCF}(C_{\infty v})$
1s	$^2\Sigma^-$		544.2	564.06[c]	556.58	542.64
	$^4\Sigma^-$		543.1		554.44	542.03
$2\sigma_g$	$^2\Sigma_g^-$		41.6	44.87	45.88	
	$^4\Sigma_g^-$		39.6		40.99	
$2\sigma_u$	$^2\Sigma_u^-$		27.9	29.90	33.45	
	$^4\Sigma_u^-$	24.5	25.3		26.04	
$3\sigma_g$	$^2\Sigma_g^-$	20.31	21.1	20.02	21.03	
	$^4\Sigma_g^-$	18.18	18.8		17.34	
$1\pi_u$	$^2\Pi_u$	16.26		19.19	16.78, etc.	
	$^4\Pi_u$		17.0		14.28	
$1\pi_g$	$^2\Pi_g$	12.10	13.1	14.47	13.09	

[a] See footnote to Table 6.

[b] ΔE_{SCF} is $E^{SCF}(O_2) - E^{SCF}(O_2^+)$ with an electron missing from the indicated orbital), assuming the usual $D_{\infty h}$ symmetry for both O_2 and O_2^+ in the case of $\Delta E^{SCF}(D_{\infty h})$, but assuming $D_{\infty h}$ for O_2 and $C_{\infty v}$ for O_2^+ in the case of $\Delta E^{SCF}(C_{\infty v})$.

[c] The $-\varepsilon$ values are 564.07 for $1\sigma_g$ and 564.05 for $1\sigma_u$—the value given here is their mean.

For O_2 the ground state is $1\sigma_g^2 1\sigma_u^2 2\sigma_g^2 2\sigma_u^2 3\sigma_g^2 1\pi_u^4 1\pi_g^2$, $^3\Sigma_g^-$. In the case of O_2, Bagus and Schaefer (BS) have made SCF calculations on the ground state of O_2 and separately on several excited "hole states" of O_2^+ corresponding to the removal of an electron from each of the several occupied MOs [13a].[3] In this way, relaxation as a cause of deviations from Koopmans' theorem [cause (a)] is removed. As can be seen from Table 7, the agreements of the resulting $\Delta E_{SCF}(D_{\infty h})$ values with the experimental I_{vert} values for the 1s electron are now somewhat improved as compared with the $-\varepsilon$ value but still leave a rather large discrepancy. [The fact that there are *two* I_{vert} values for the K electron is explained by exchange coupling of the π_g^2, $^3\Sigma_g^-$ electron pair in O_2 with the 1s shell to give $^2\Sigma^-$ and $^4\Sigma^-$ ions after one 1s electron is removed; this is taken into account in the calculations, but the difference (see footnote *c* of Table 8) between the calculated ε values for $1\sigma_g$ and $1\sigma_u$, too small to be detected in the experimental I_{vert}, is ignored and we put $1\sigma_g$ and $1\sigma_u$ together as 1s.]

In seeking to explain the remaining discrepancy, BS were not satisfied to attribute it to cause (b), but tried the assumption that O_2^+ in a 1s hole

[3] Also Bagus *et al.* [13b]; and for earlier work see Davis *et al.* [13c].

state acts like a heteropolar molecule (symmetry $C_{\infty v}$) such as FO. Although the final overall symmetry of the wave function must correspond to $D_{\infty h}$, this result can be secured by making two linear combinations, differing only a little in energy, of the two $C_{\infty v}$ K-hole wave functions corresponding to O^+O and OO^+: $\Psi(D_{\infty h}) = \Psi(O^+O) \pm \Psi(OO^+)$. BS hence made a $C_{\infty v}$ SCF calculation for $O_2{}^+$; the difference between the resulting computed SCF energy and that of neutral O_2 is given as $\Delta E_{SCF}(C_{\infty v})$ in Table 8. It is

TABLE 8

Gross Populations on Centers A and B in $\Psi(O^+O)$

Shell	Center A	Center B
$1s_A$	1.00	0.00
$1s_B$	0.000	2.00
$\sim 2\sigma_g$	1.13	0.87
$\sim 2\sigma_u$	0.92	1.08
$\sim 3\sigma_g$	1.03	0.97
$\sim 1\pi_u$	3.44	0.56
$\sim 1\pi_g$	0.26	1.74
Totals:	7.78	7.22

seen that this time there is good agreement with the experimental I_{vert} values. The remaining small discrepancy can be attributed to differences in correlation (plus relativistic) energy between the O_2 and $O_2{}^+$ states involved.

Further, BS have considered the effect on the valence electron shells of the O^+O or OO^+ polarization in $O_2{}^+$ in a 1s hole state. They present their result in terms of a population analysis of the gross atomic populations on the two centers in $\Psi(O^+O)$. Table 8 reproduces their results. These show that the complete O^+O polarization in the core causes a strong reactive polarization in the outer shells, with the π shells being much the most polarizable. The LCAO-*additive* MOs ($2\sigma_g$, $3\sigma_g$, $1\pi_u$) are polarized *toward*, but the subtractive ones ($2\sigma_u$, $1\pi_g$) *away from*, the O^+ center. The MOs are all heteropolar as for $C_{\infty v}$ but retain some resemblance to their homopolar forms in neutral O_2.

The occurrence of *localized* instead of delocalized 1s hole states in $O_2{}^+$ clearly indicates that like effects must occur whenever a K shell is ionized in a molecule containing atoms in equivalent locations, as in Li_2, N_2, F_2, C_2H_2, C_2H_4, N_2H_4, N_2O_4, BF_3, CF_4, etc. They indicate further that in such cases a third cause (c) must be added to the two (a and b) already given for the failure of exact agreement between I_{vert} values and the $-\varepsilon$ values of a neutral molecule. In particular, this conclusion applies to Table 6 for N_2. In the case of molecules with several inner shells, for example P_2, similar effects may be expected, *to a greater or lesser extent, for all the core shells*, not just for the K shell. Similar effects are expected also for excitation energies [14].

The results obtained by BS in departing from our usual procedure of using only MOs having the overall symmetry of the molecule (restricted Hartree–Fock SCF method) could also be obtained by suitable electron correlation. They are an example of (partially) "unrestricted Hartree–Fock" methods.

C. MOMENTUM WAVE FUNCTIONS AND MOMENTUM DENSITIES

Our discussion is almost entirely in terms of wave functions in *position* space, but wave functions in momentum space are also of interest. These can be obtained by a Dirac transformation to go from the usual MOs to MOs in momentum space. Coulson and Duncanson [15a][4] discussed this problem extensively some time ago, but did not have good wave functions except for H_2^+ and H_2. Recently Cade and Henneker have used the good SCF functions reviewed in Section A to obtain corresponding good momentum wave functions for N_2, O_2, and LiF [16], and from these, direct and difference density functions in momentum space, analogous to those in position space discussed in Section C. The most obvious result in the molecules mentioned is a general *decrease* in the probability of finding an electron with low momentum and an increase of the probability of finding it with higher momentum in the molecule as compared with the separate atoms. In particular, the component of electronic momentum perpendicular to the symmetry axis is notably increased in molecule formation.

In another paper, these results are compared with derived predictions concerning Compton scattering of X rays from N_2 and O_2 [17]. The theoretical and experimental Compton profiles are in good agreement when good SCF wave functions are used. A paper by Tawil and Langhoff [18] discuss calculated Compton profiles in N_2 and O_2, both by SCF and with CM; the agreement with experiment is appreciably improved by CM.

D. CHARGE DISTRIBUTIONS

Contour maps for SCF electron densities have been discussed for diatomic hydrides in Section IV.B, to which reference should be made. A similar discussion applies here to the contour maps for first-row homopolar

[4] For a recent detailed review see Epstein [15b].

diatomic molecules shown in Figs. 2 and 3, taken from a paper by Bader *et al.* [19] and based on the SCF functions considered in Section A, all at the R_e values of the respective molecules. Figure 2 shows the overall densities in atomic units at a cross section through the symmetry axis, and Fig. 3 shows the corresponding difference densities (molecular densities minus the superposed densities of the separate atoms). Figures 4–10, from a paper by Wahl [20], show contour diagrams for the several MOs; the contours are spaced so that each new contour, going from outside in, represents an increase of 2 in the density. The outermost contour corresponds to a density of 6.1×10^{-5} a.u. Boyd gives a total density and difference density map for P_2 [21].

Kohl and Bartell compare SCF computed densities with corresponding experimental data obtained from electron diffraction in several molecules [22].

A striking feature of the valence-shell MOs is the considerable resemblance of their forms to those of AOs of a single atom placed at the center of the molecule. In N_2, $2\sigma_g$ resembles an atomic 2s AO except for the presence of a *double* nodal surface (one around each nucleus); $2\sigma_u$ resembles a $2p\sigma$ AO except for the *addition* of a radial node around each nucleus; $1\pi_u$ is a somewhat distorted $2p\pi$; $3\sigma_g$ shows a pronounced resemblance to a high-ζ $3d\sigma$; and $1\pi_g$ is very similar to a high-ζ $3d\pi$. These resemblances have been summarized by describing the molecule as a semi-united atom [23]. Langmuir has described the N_2 molecule ($1\sigma_g{}^2 1\sigma_u{}^2 2\sigma_g{}^2 2\sigma_u{}^2 1\pi_u{}^4 3\sigma_g{}^2$) as having a valence-shell octet like that of a Ne atom ($2s^2 2p^6$, corresponding to $2\sigma_g{}^2 2\sigma_u{}^2 1\pi_u{}^4$ in N_2) plus an "imprisoned pair" ($3\sigma_g{}^2$) outside the two K shells of the two N atoms [23]. Except for the two K shells and the substitution of a tightly bound 3d-like $3\sigma_g$ orbital pair instead of a more loosely bound 3s pair, N_2 considerably resembles a Mg atom. By expanding the MOs in terms of single-center functions, Huzinaga has shown [23] that $2\sigma_g$ has a large component of single-center s character (plus some $3d\sigma$), $2\sigma_u$ is largely $2p\sigma$, $1\pi_u$ is largely $2p\pi$, and $3\sigma_g$ is mainly $3d\sigma$ (plus some 2s).

E. POPULATION ANALYSIS AND BONDING

Considerations here on overlap population (Tables 9–12) and chemical bonding (Table 13) are similar to those in Chapter IV (q.v.). In the first-row molecules there is major s, $p\sigma$ hybridization in the $2\sigma_g$, $2\sigma_u$, and $3\sigma_g$ MOs, while $d\pi$ (plus some $f\pi$), participation is important in strengthening the bonding in $1\pi_u$. The positive s, $p\sigma$ hybridization in $2\sigma_g$ makes this MO very strongly bonding, while negative hybridization in $3\sigma_g$ makes it only weakly bonding. In N_2, judging from the overlap population, the negative hybridization in $2\sigma_u$ makes this MO weakly bonding instead of antibonding. However, the fact that ω_e is somewhat increased and R_e slightly decreased in

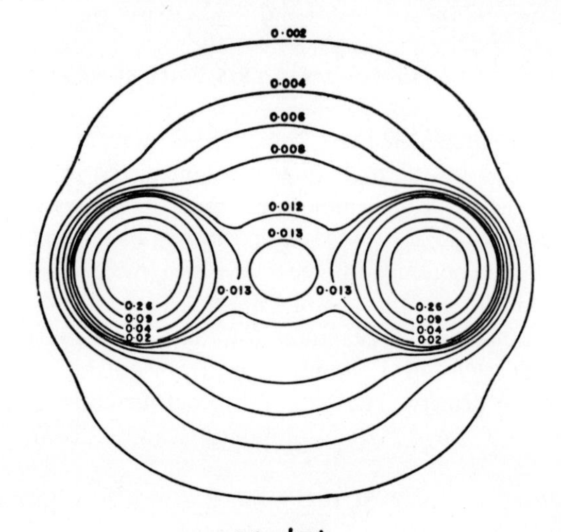

$Li_2 \; ^1\Sigma_g^+$

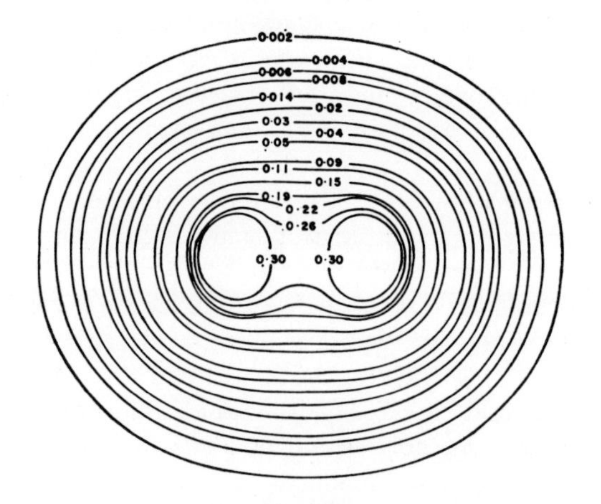

$C_2 \; ^1\Sigma_g^+$

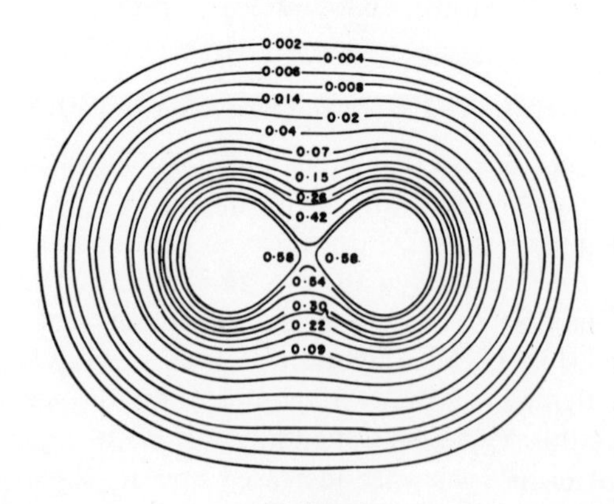

$O_2 \; ^3\Sigma_g^-$

FIG. 2 Total molecular charge density contours (a.u.). [From R. F. W. Bader, W. H. Henneker, and P. E. Cade, *J. Chem. Phys*. **46**, 1207 (1972).]

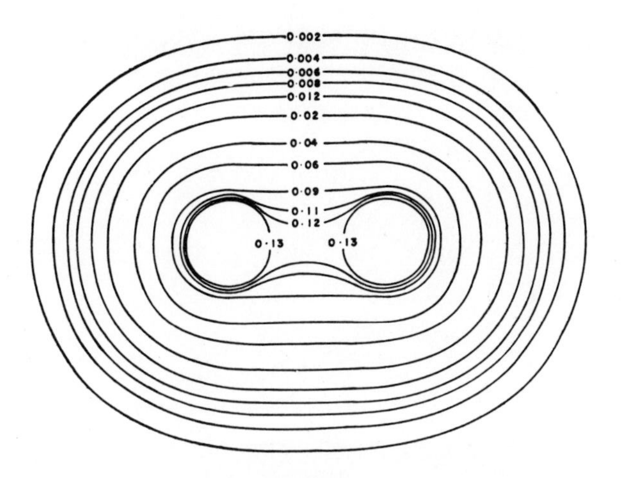

$B_2 \ {}^3\Sigma_g^-$

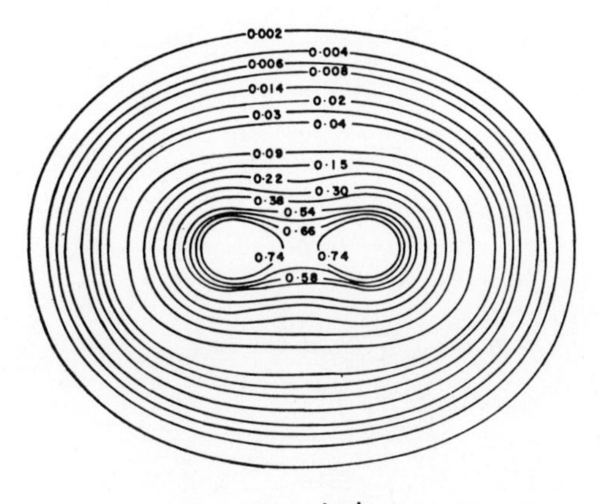

$N_2 \ {}^1\Sigma_g^+$

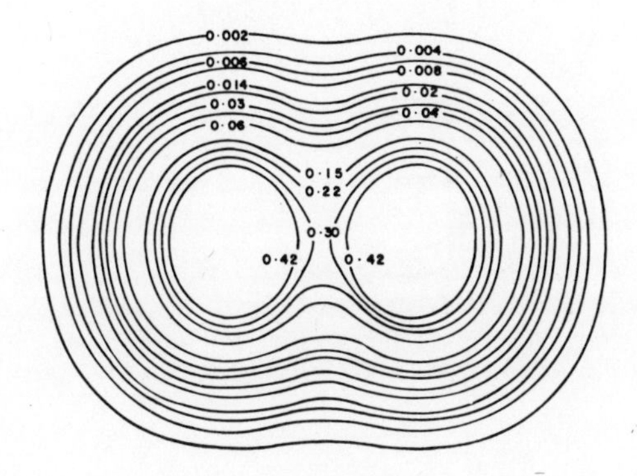

$F_2 \ {}^1\Sigma_g^+$

FIG. 2 (*continued*)

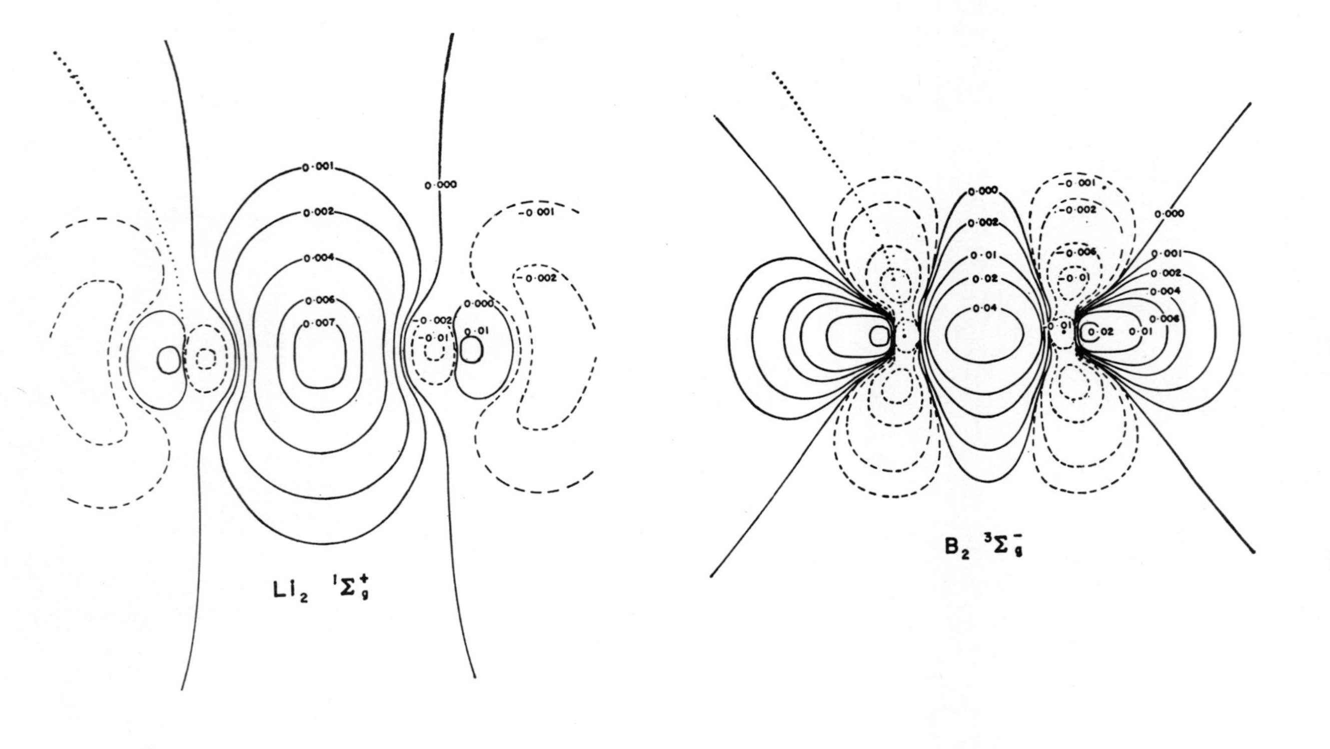

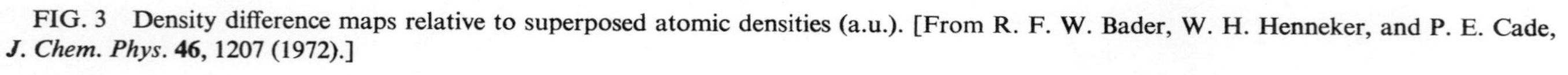
FIG. 3 Density difference maps relative to superposed atomic densities (a.u.). [From R. F. W. Bader, W. H. Henneker, and P. E. Cade, *J. Chem. Phys.* **46**, 1207 (1972).]

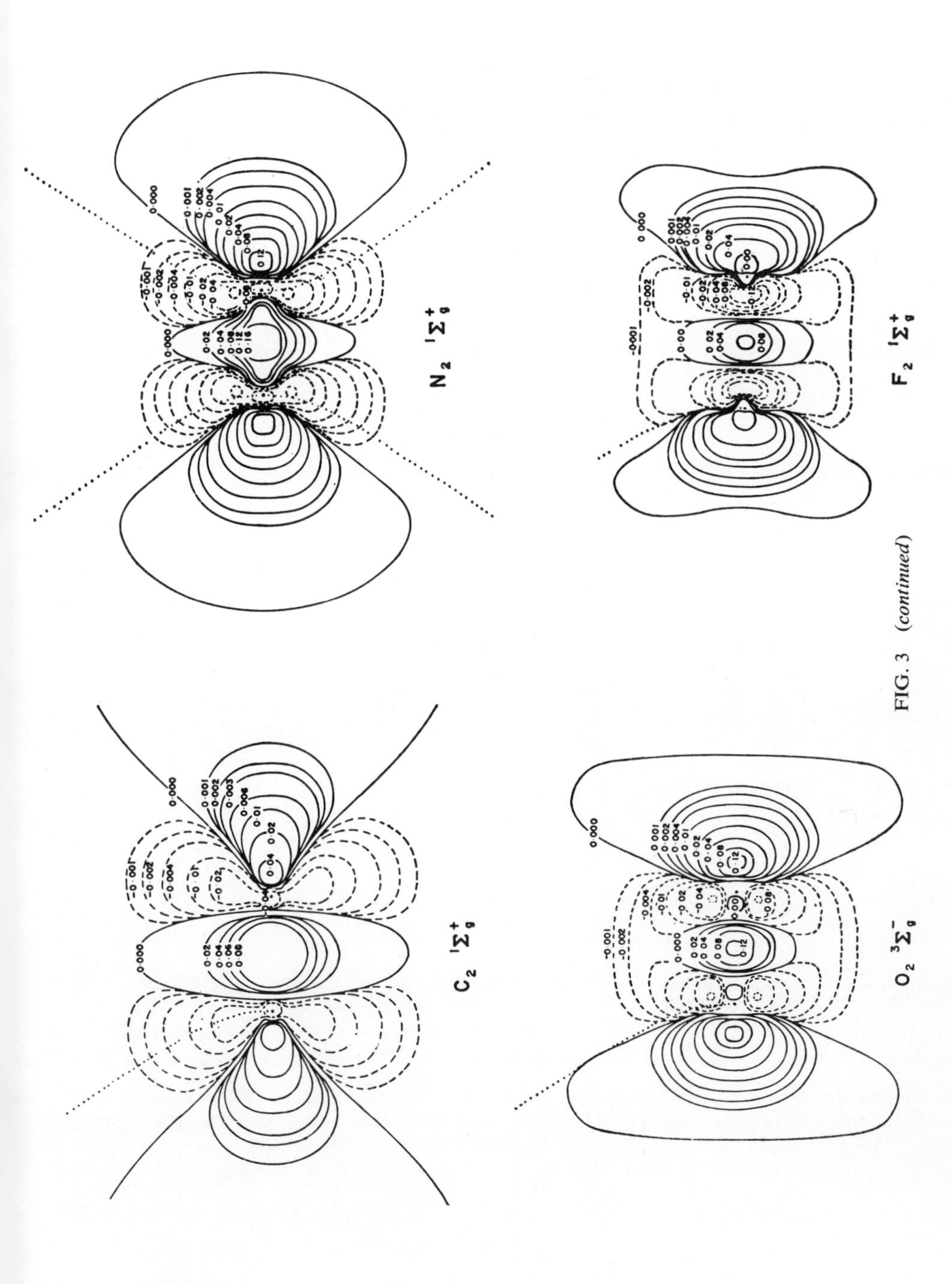

FIG. 3 *(continued)*

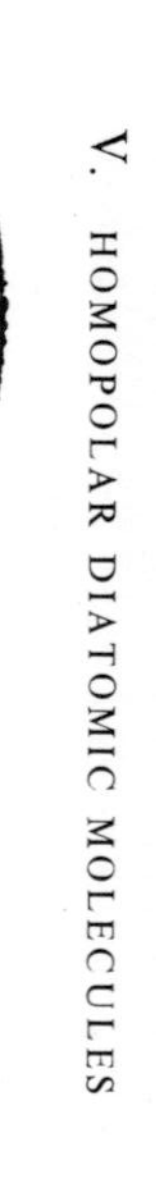

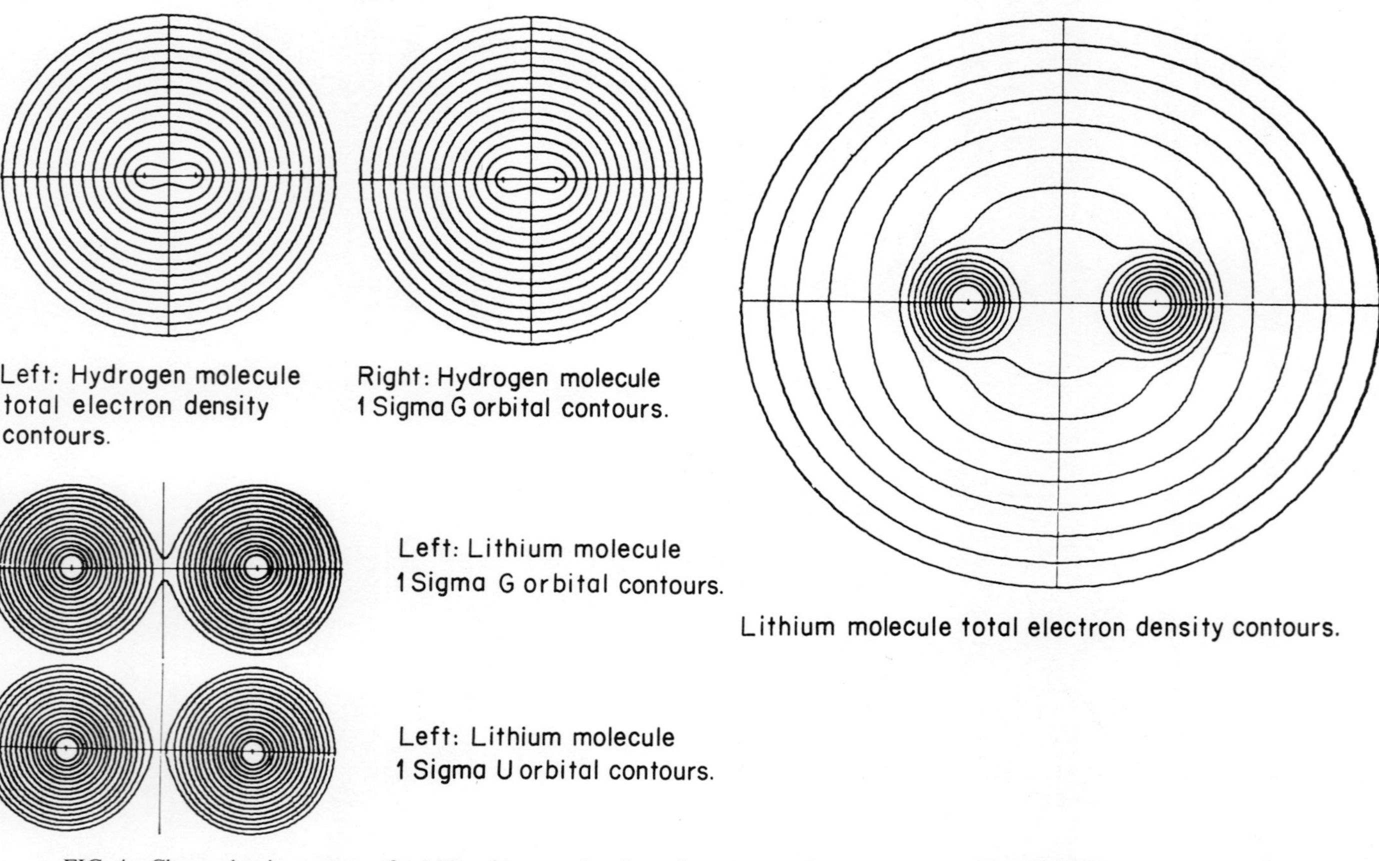

FIG. 4 Charge density contours for MOs of homopolar diatomic molecules. [From A. C. Wahl, *Science* **151**, 961 (1966). Copyright 1966 by the American Association for the Advancement of Science.]

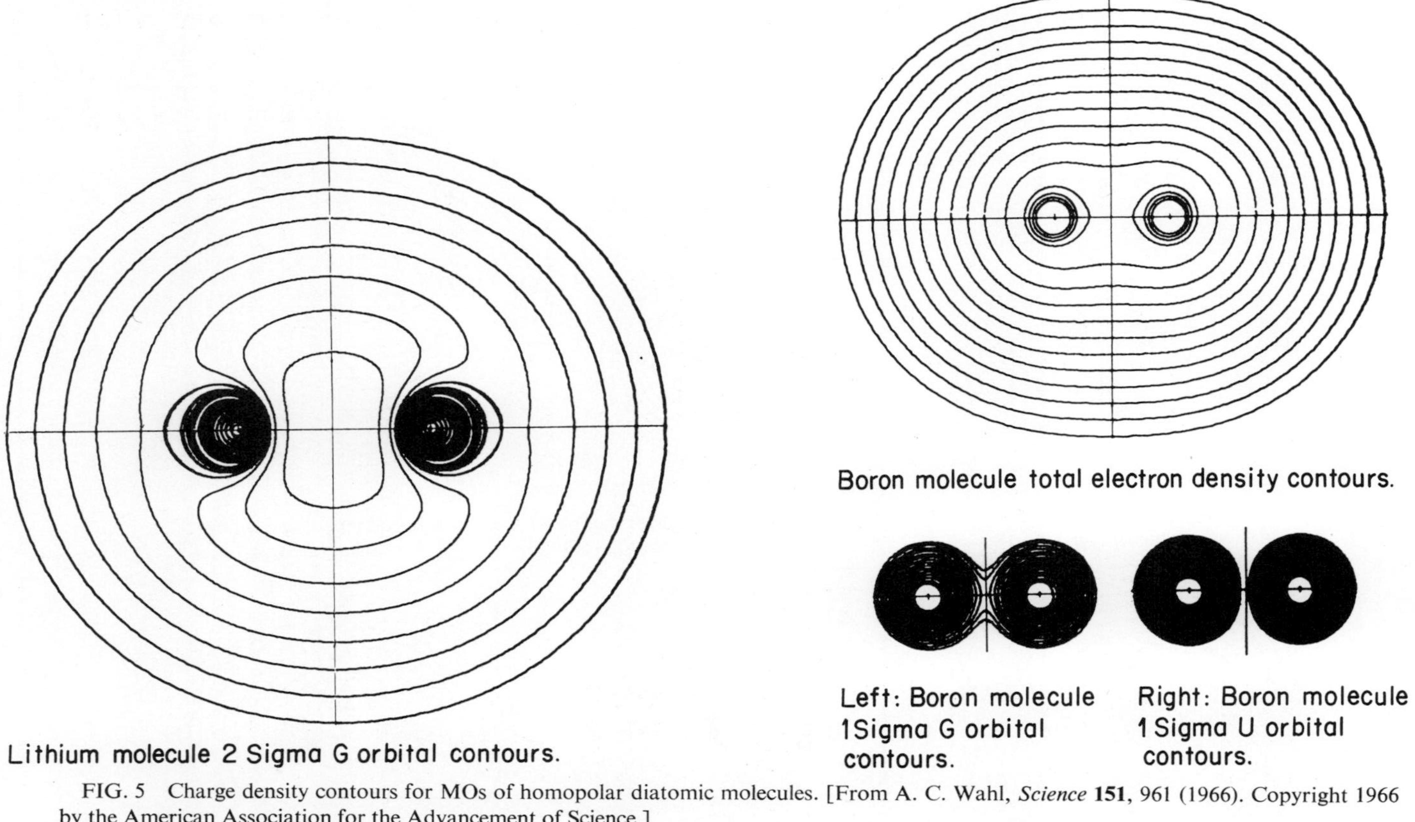

FIG. 5 Charge density contours for MOs of homopolar diatomic molecules. [From A. C. Wahl, *Science* **151**, 961 (1966). Copyright 1966 by the American Association for the Advancement of Science.]

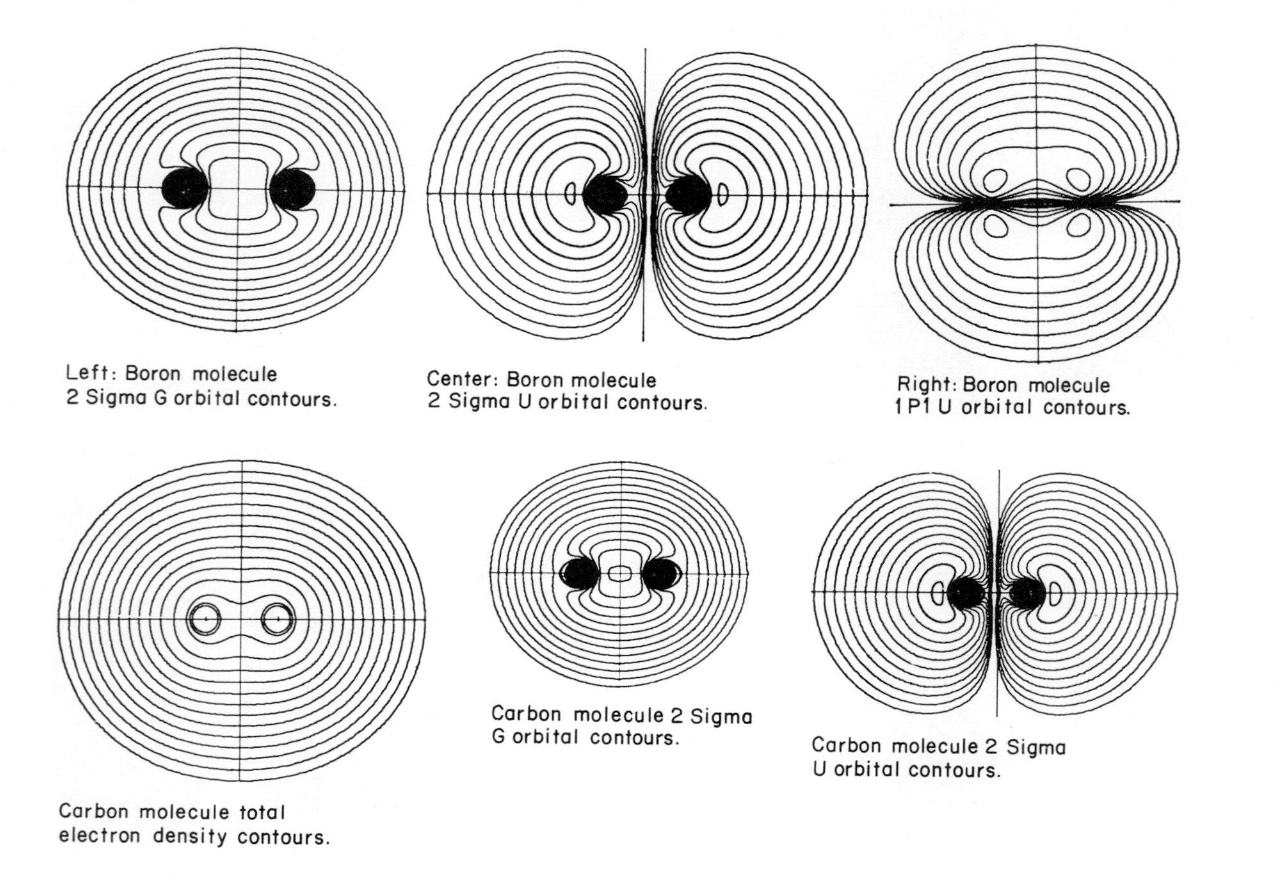

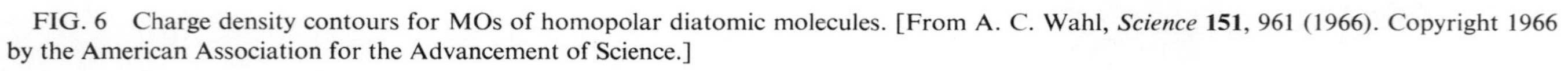
FIG. 6 Charge density contours for MOs of homopolar diatomic molecules. [From A. C. Wahl, *Science* **151**, 961 (1966). Copyright 1966 by the American Association for the Advancement of Science.]

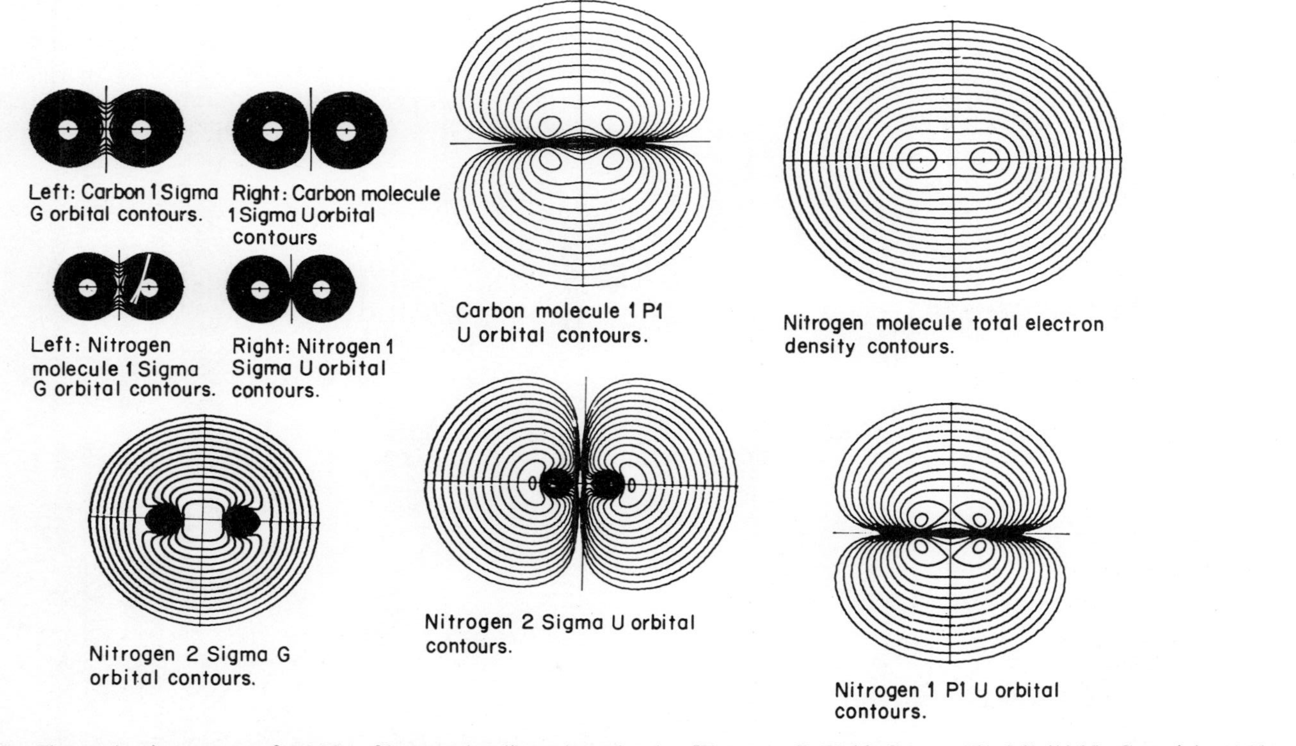

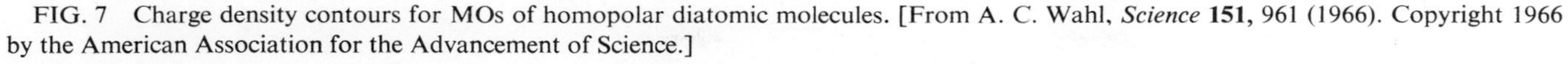
FIG. 7 Charge density contours for MOs of homopolar diatomic molecules. [From A. C. Wahl, *Science* **151**, 961 (1966). Copyright 1966 by the American Association for the Advancement of Science.]

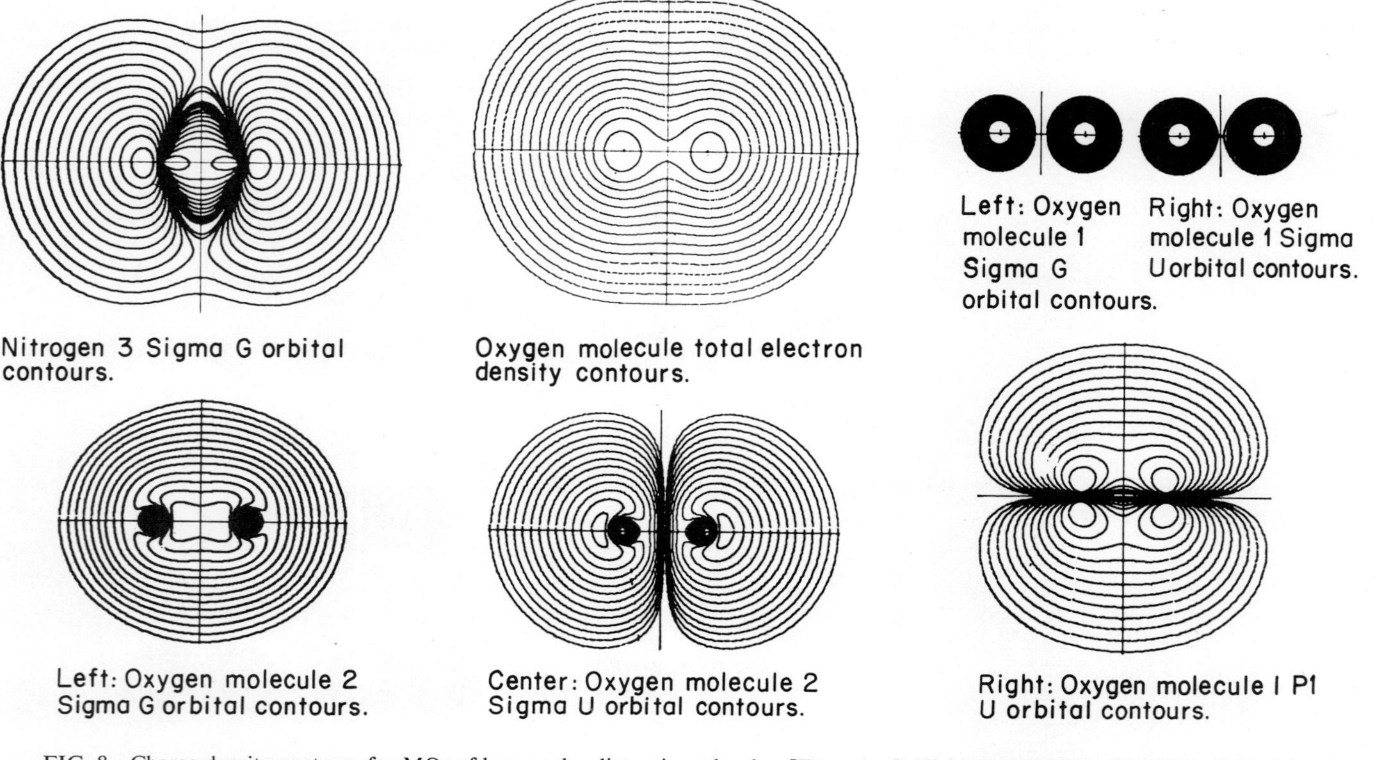

FIG. 8 Charge density contours for MOs of homopolar diatomic molecules. [From A. C. Wahl, *Science* **151**, 961 (1966). Copyright 1966 by the American Association for the Advancement of Science.]

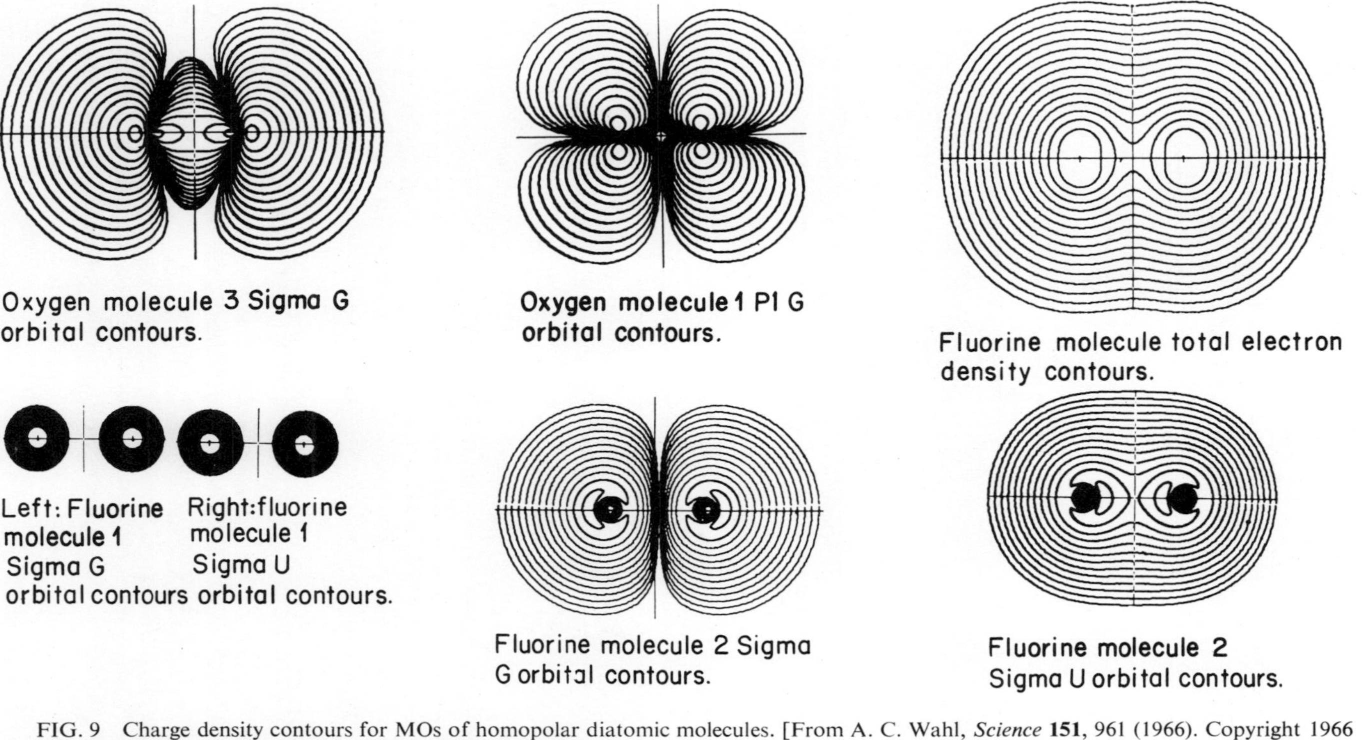

FIG. 9 Charge density contours for MOs of homopolar diatomic molecules. [From A. C. Wahl, *Science* **151**, 961 (1966). Copyright 1966 by the American Association for the Advancement of Science.]

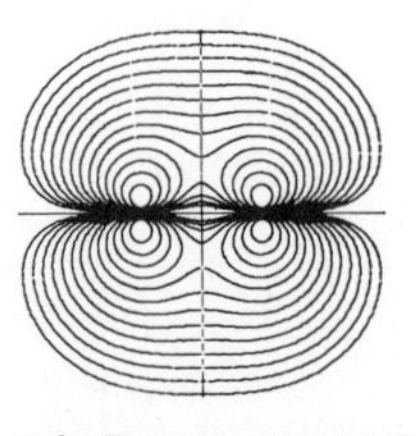

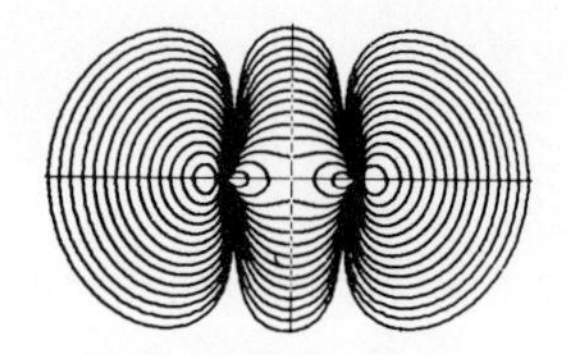

FIG. 10 Charge density contours for MOs of homopolar diatomic molecules. [From A. C. Wahl, *Science* **151**, 961 (1966). Copyright 1966 by the American Association for the Advancement of Science.]

TABLE 9

Overlap Populations for Li_2 [a]

$MO(\phi_i)$	n_i(s, s)	n_i(s, pσ)	n_i(pσ, pσ)	n_i(dσ, all)	$n_i(Li_2)$
$1\sigma_g$	0.001	0.000	−0.000	0.000	0.001
$1\sigma_u$	−0.001	−0.000	0.000		−0.001
$2\sigma_g$	0.617	0.197	0.015	0.005	0.833
Totals:	0.616	0.197	0.015	0.005	0.833

[a] n_i(s, s) is the sum of all inter-s STF overlaps; n_i(s, pσ) of all overlaps of s with pσ STFs; n_i(pσ, pσ) of all inter-pσ STF overlaps; n_i(dσ, all) is the sum of all overlaps of dσ STFs with s, pσ, and dσ STFs; $n_i(Li_2)$ is the total overlap population for ϕ_i in Li_2.

N_2^+ after removal of a $2\sigma_u$ electron shows that $2\sigma_u$ after all is still distinctly antibonding.

In a simplistic discussion, one might have expected the net effect of $2\sigma_g^2$ (roughly $\sigma_g 2s^2$) and $2\sigma_u^2$ (roughly $\sigma_u 2s^2$) to be a net repulsion as in the case of $1\sigma_g^2 1\sigma_u^2$ (equally well describable as $1s^2 1s^2$) in the He_2 molecule. However, "free" s, pσ hybridization more or less removes the antibonding effect in $2\sigma_u$, which would be expected if it were pure $\sigma_u 2s$. On the other hand, the bonding which might have been expected if $3\sigma_g$ were pure $\sigma_g 2p$ is largely transferred to $2\sigma_g$ *as a result of "forced" hybridization.* A more detailed

TABLE 10

Overlap Populations for N_2[a]

ϕ_i	$n_i(s, s)$	$n_i(s, p\sigma)$	$n_i(p\sigma, p\sigma)$	$n_i(df\sigma, all)$	$n_i(p\pi, p\pi)$	$n_i(df\pi, all)$	$n_i(\sigma)$	$n_i(\pi)$	$n_i(N_2)$
$1\sigma_g$	0.001	0.000	0.000	0.001			0.002		0.002
$1\sigma_u$	0.002	0.003	−0.000	0.001			0.006		0.006
$2\sigma_g$	0.333	0.374	0.067	0.072			0.847		0.847
$2\sigma_u$	−0.594	0.789	−0.100	0.052			0.148		0.148
$1\pi_u$					0.915	0.241		1.156	1.156
$3\sigma_g$	0.414	−0.916	0.335	−0.070			−0.237		0.006
Totals:	0.156	0.252	0.302	0.055	0.915	0.241	0.766	1.156	1.922

[a] Cf. footnote to Table 9. But here n_i (σ) and n_i (π) are the total σ and π overlaps for ϕ_i, while $n_i(N_2)$ is the overall total for ϕ_i.

TABLE 11

Overlap Populations for F_2[a]

ϕ_i	n_i(s, s)	n_i(s, pσ)	n_i(pσ, pσ)	n_i(dfσ, all)	n_i(pπ, pπ)	n_i(dfπ, all)	$n_i(\sigma)$	$n_i(\pi)$	$n_i(F_2)$
$1\sigma_g$	0.000	0.000	0.000	0.000			0.000		0.000
$1\sigma_u$	0.000	0.000	−0.000	0.000			0.000		0.000
$2\sigma_g$	0.214	0.089	0.006	0.047			0.356		0.356
$2\sigma_u$	−0.304	0.103	−0.006	0.011			−0.197		−0.197
$3\sigma_g$	0.039	−0.295	0.364	0.012			0.120		0.120
$1\pi_u$					0.376	0.062		0.438	0.438
$1\pi_g$					−0.516	0.005		−0.511	−0.511
Totals:	−0.052	−0.102	0.363	0.070	−0.140	0.067	0.280	−0.073	0.206

[a] Cf. footnote to Table 10.

TABLE 12

Overlap Population for Valence Shells of P_2[a]

ϕ_i	n_i(s, s)	n_i(s, pσ)	n_i(pσ, pσ)	n_i(dfσ, all)	n_i(pπ, pπ)	n_i(dfπ, all)	$n_i(\sigma)$	$n_i(\pi)$	$n_i(P_2)$
$4\sigma_g$	0.356	0.225	0.032	0.079			0.693		0.693
$4\sigma_u$	−0.466	0.618	−0.084	0.035			0.103		0.103
$5\sigma_g$	0.216	−0.664	0.491	0.044			0.087		0.087
$2\pi_u$					0.833	0.385		1.218	1.218
Totals:	0.106	0.179	0.439	0.158	0.833	0.385	0.883	1.218	2.101

[a] Cf. footnote to Table 10. Note that $n_i(P_2)$ would be slightly decreased if overlaps involving the inner shells were included.

TABLE 13

Overlap and Bonding in First-Row Homopolar Molecules and P_2

Molecule	n	R_e (a.u.)	I (eV)	$n\bar{I}$ (eV)	D_e (eV)
Li_2	0.833	5.050	5.39	4.49	1.10
C_2	1.557	2.347	11.3	17.6	4.9
N_2	1.922	2.074	14.5	27.9	9.76
O_2	0.721	2.282	13.6	9.80	5.08
F_2	0.206	2.68	17.4	3.58	1.6
P_2	2.10	3.497	11.0	23.1	5.03

discussion of forced and free hybridization in N_2 is given by Mulliken [24]. The net effect of the bonding in the three σ MOs is equivalent to one strong σ bond. The strength of the s–pσ hybridization which makes this bond so effective in N_2 is due in part to bond shortening by the presence of the bonding $1\pi_u^4$ shell, yielding altogether a strong triple bond, which gives to N_2 an exceptionally small R_e and large D.

It is notable that, as in the diatomic hydrides, the σ bonding is largely concentrated in the lowest-energy valence-shell σ MO, here the $2\sigma_g$. (However, see the discussion in Section IV.C of what happens when MCSCF instead of SCF MOs are used.) This is least true in the weakly bound F_2 molecule, where all hybridizations are much reduced, so that $2\sigma_g^2 2\sigma_u^2$ tends to resemble $1\sigma_g^2 1\sigma_u^2$ in He_2, while $3\sigma_g^2$ becomes a definitely, although weakly, bonding MO (note the overlap populations in Table 11 for F_2 compared with those for N_2 in Table 9). At the same time, $1\pi_u^4 1\pi_g^4$ destabilizes F_2 in a similar way to $1\sigma_g^2 1\sigma_u^2$ in He_2, although dπ mixing, by somewhat increasing the bonding in $1\pi_u$, acts in the opposite direction.

As with the diatomic hydrides, the products nI, where n is the total overlap population and I the ionization potential, are roughly proportional to D (see Table 13).

From the gross atomic populations in Table 14–17, one can obtain the

TABLE 14

Gross Atomic Populations for Li_2

ϕ_i	N_i(s)	N_i(pσ)	N_i(dσ)	N_i(Li_2)
$1\sigma_g$	2.000	0.000	0.000	2.00
$1\sigma_u$	2.000	−0.000		2.00
$2\sigma_g$	1.833	0.164	0.003	2.00
Totals:	5.833	0.164	0.003	6.00

TABLE 15

Gross Atomic Populations for N_2

ϕ_i	N_i(s)	N_i(pσ)	N_i(dσ)	N_i(fσ)	N_i(pπ)	N_i(dπ)	N_i(fπ)	N_i(N_2)
$1\sigma_g$	1.999	0.000	0.000	0.000				2.000
$1\sigma_u$	1.998	0.002	0.000					2.000
$2\sigma_g$	1.439	0.515	0.042	0.004				2.000
$2\sigma_u$	1.191	0.780	0.029					2.000
$1\pi_u$					3.851	0.135	0.013	4.000
$3\sigma_g$	0.657	1.368	−0.023	−0.002				2.000
Totals:	7.285	2.665	0.048	0.002	3.851	0.135	0.013	14.000

TABLE 16

Gross Atomic Population for F_2

ϕ_i	N_i(s)	N_i(pσ)	N_i(dσ)	N_i(fσ)	N_i(pπ)	N_i(dπ)	N_i(fπ)	N_i(N_2)
$1\sigma_g$	2.000	0.000	−0.000	0.000				2.000
$1\sigma_u$	2.000	0.000	0.000					2.000
$2\sigma_g$	1.895	0.080	0.019	0.006				2.000
$2\sigma_u$	1.921	0.073	0.006					2.000
$3\sigma_g$	0.086	1.899	0.013	0.003				2.000
$1\pi_u$					3.966	0.022	0.012	4.000
$1\pi_g$					3.998	0.005	−0.003	4.000
Totals:	7.901	2.052	0.038	0.009	7.964	0.027	0.009	18.000

TABLE 17

Gross Atomic Populations of Valence-Shell MOs for P_2 at R_e [a]

ϕ_i	N_i(s)	N_i(pσ)	N_i(dσ)	N_i(fσ)	N_i(pπ)	N_i(dπ)	N_i(fπ)	$N_i(P_2)$
$4\sigma_g$	1.70 (1.69)	0.25 (0.31)	0.046	0.005				2.000
$4\sigma_u$	1.47 (1.78)	0.51 (0.22)	0.019	0.001				2.000
$5\sigma_g$	0.34 (0.43)	1.63 (1.57)	0.033	0.000				2.000
$2\pi_u$					3.73 (4.00)	0.240	0.032	4.000
Totals:	3.51	2.39	0.099	0.006	3.73	0.240	0.032	10.000

[a] The figures in parentheses refer to calculations in which the *d* and *f* functions were deleted.

extent of s–pσ hybridization or promotion, also of d orbital participation (cf. Section IV.C). Note how the d population is enhanced, especially for the π_u MOs, in P_2 as compared with N_2, although it is surprising that the increase is not greater.

The results for P_2 in Tables 12, 13, and 17 are based on accurate SCF calculations by Mulliken and Liu [25]. In this same paper, the role of d functions in chemical bonding is discussed. Also, computations are made there which show the importance of free hybridization in the $2\sigma_u$ MO of N_2: if this MO were pure $\sigma_u 2s$, the total energy of N_2 would be raised by 0.123 a.u. or 3.37 eV. Thus the relief of antibonding by s, p hybridization in $2\sigma_u$ contributes very appreciably to the bond strength in N_2. An analogous but smaller effect (0.021 a.u.) is found in the $4\sigma_u$ MO of P_2.

F. ELECTRON CORRELATION

A number of papers have been published on wave functions of first-row homopolar diatomic molecules including varying degrees of CM to secure electron correlation.[5] On the general subject, reference should first be made to the discussion in Sections IV.D and IV.E. Among homopolar diatomic molecules, C_2 shows an exceptionally large CM in its ground state at R_e (see Section III.D). The same is true of N_2^+.

In some papers of Das and Wahl [27], the OVC method (see Section IV.D) is used to obtain wave functions for Li_2 and F_2 which fairly correctly give the forms of molecular potential curves out to dissociation; further improvements by additional CM are also given.

A promising systematic procedure for dealing with the problem of electron correlation has been described in a recent paper by Buenker and Peyerimhoff [28] somewhat as follows.

(1) Choose an adequate basis set (of STFs, GTFs, GLFs, or possibly one using elliptical coordinates).

(2) Choose a small number of *main* configurations to form a ψ_0 as the nucleus of a CM treatment; "experience has shown that it is generally sufficient to include only configurations with a coefficient 0.05 or greater" in the final CM. (The 0.05 threshold applies at whatever R value the coefficient is largest if one wishes to go out to dissociation, and also for excited states of the same species if one wishes to compute these at the same time by using higher roots of the same secular equation.) For this ψ_0 one carries out an MCSCF treatment.

(3) From ψ_0 form all configurations which can be obtained by single or

[5] For a list including other calculations through 1969, see Richards *et al.* [26].

double excitations into the virtual MOs of the original MCSCF treatment. Test the importance of these one at a time by solving a secular equation including ψ_0 and the configuration to be tested, say the rth configuration, and determine the energy increment ΔE_r which results. (Or alternatively, use second-order perturbation theory to test the importance of each added configuration; however, Buenker and Peyerimhoff prefer the secular equation procedure as economical in the long run.) Choose a restricted number of configurations including all those whose ΔE_r's exceed a certain threshold value T (conveniently expressed in microhartrees, where 1 hartree is 1 a.u.).

(4) Using a relatively large T, carry out a CM calculation and then convert the MOs to corresponding NOs by diagonalizing the density matrix for the state of interest (see Chapter I).

(5) Perhaps revise ψ_0 to include fewer main configurations, as very often becomes possible when NOs are used.

(6) Solve the CM secular problem at each of several levels of T, including ψ_0 and all the singly and doubly excited configurations derived from it using the NOs at that T level.

(7) At each T level, extrapolate to $T = 0$ by adding to $E(T)$ the sum of the ΔE_r's for all configurations *not* yet included.

(8) In general, the extrapolated $E(T \to 0)$ values from the different T levels do not agree, but increase monotonically; by plotting $E(T \to 0)$ against T, a fairly good value of $E(T = 0)$ may be obtained.

In this procedure, the effects of triple and higher excitations are probably negligible if enough main configurations from which excitations are made are included in ψ_0. The procedure is applicable to polyatomic as well as diatomic molecules. In the interest of practicality, excitations from some of the inner MO shells may have to be omitted. The full implementation of a procedure such as that just outlined is obviously time-consuming. Most of the work so far has not gone beyond stage (5) at most.

Before discussing some of the published papers, we consider a maximally simplified procedure which has proved useful in predicting numerous excited-state potential curves of the molecules considered. The method, which Schaefer and Harris [29] call the VCI (valence CM) method, is as follows. With a *minimal* basis set, solve a CM problem including all CSFs which can be obtained by adding electron excitations from the SCF function into the valence-shell MOs which are unoccupied in the latter. For example, for N_2 (SCF configuration $1\sigma_g^2 1\sigma_u^2 2\sigma_g^2 2\sigma_u^2 1\pi_u^2 3\sigma_g^2$), the VCI excitations are into $1\pi_g$ and $3\sigma_u$. However, excitation out of the K shells ($1\sigma_g^2 1\sigma_u^2$) is usually omitted. The specified VCI excitations are adequate to provide a CM wave function which dissociates properly, that is, leads as $R \to \infty$ to atoms in combinations of the correct valence-shell states (e.g., for N_2 in combinations of $1s^2 2s^2 2p^3$, ^4S, ^2D, and ^2P states). By obtaining the CM coefficients and energies E for each of a number of R values, the potential curve for any desired valence-shell shell state can be computed to a surprisingly good approximation by plotting $E(R)$. The curves can be further adjusted semiempirically by comparisons with those that are known from empirical spectroscopic evidence.

Using this method, Schaeffer and Harris [29] have obtained approximate potential curves for all 62 valence-shell states of O_2. Similarly, Michels has computed potential curves for all the 102 valence-shell excited states of N_2, and by calibrating against known excited states, has estimated the true forms of the potential curves. Of particular interest are two predicted low-energy states, a $^1\Sigma_g{}^+$ and a $^3\Delta_g$ (of which only the $^3\Delta_g$ has been observed), whose main configuration is $\ldots 1\pi_u{}^2 3\sigma_g{}^2 1\pi_g{}^2$; the $^1\Sigma_g{}^+$ is the lowest excited state of its species, with a dissociation energy of nearly 4 eV [30].[6]

Thulstrup and Andersen have made similar minimal-basis plus CM calculations [31] and obtained potential curves for numerous known and predicted low-lying states of $N_2{}^-$, $N_2{}^+$, and N_2^{2+}. Using a less limited basis set plus CM, Cartwright and Dunning [32] have also computed potential curves for numerous states of $N_2{}^+$.

Earlier, Fougere and Nesbet in a similar way obtained potential curves for 27 states of C_2 [33]. Even in their best calculations, however, they kept the $2\sigma_g{}^2$ shell as well as the K shells closed. Similarly, but using a basis set of elliptical orbitals, Bender and Davidson [34] did a CM calculation on potential curves for nine states of Be_2; however, they computed only the lowest-energy curve of each electronic state species. The excited states all showed stable minima, but the ground state was essentially repulsive. The same authors [35a], again using an elliptical-coordinate basis set and CM, obtained potential curves for the 81 lowest-energy valence shell states of B_2. They concluded that the ground state is a stable $^5\Sigma_u{}^-$ of electron configuration $1\sigma_g{}^2 1\sigma_u{}^2 2\sigma_g{}^2 2\sigma_u 3\sigma_g 1\pi_u{}^2$, but now [35b] $a^3\Sigma_g{}^-$ state is predicted.

Several more accurate CM calculations have been made on N_2 and O_2. An instructive CM calculation for the ground state of N_2 using a contracted Gaussian-lobe basis set capable of giving 63% of the correlation energy with a single determinant Φ_0 has been described by Langhoff and Davidson [36]. They compared the results with second-order perturbation theory which overestimated the correlation energy by 23 to 50% depending on how H^0 was chosen. Pair–pair interaction affects the results by almost 20%, while quadruple excitations have an 8% effect.

Schaefer obtained a good potential curve for the ground state of O_2 from

[6] As yet unpublished in 1977 except for the note on the low $^1\Sigma_g{}^+$ and $^3\Delta_g$ states. See also Mulliken [24].

moderately accurate calculations (calculated D 4.72 eV as compared with the experimental 5.21 eV; good agreements for spectroscopic constants) [37]. Schaefer and Miller [38] used extensive CM in computations on the lowest $^3\Pi_u$ and $^3\Sigma_u^-$ states. Morokuma and Konishi [39] made calculations on several states including the lowest $^3\Sigma_u^-$ (see Section H). Peyerimhoff and Buenker [40] dealt with the three lowest states at R_e. Moss and Goddard made GVB CM computations on several states [41]. Julienne, Neumann, and Krauss using MCSCF wave functions computed vertical transition moments for the B–X and lowest $^3\Pi_u$–X transitions [42]. Further calculations of increasing accuracy on several excited states involve mixing with Rydberg states. These will be discussed in Section H.

Although for free O_2^- only the ground state is stable, there is experimental evidence of some excited states in crystals and also of unstable excited states of the free ion. Krauss *et al.* [43] have made MCSCF and CM calculations from which they have obtained a full set of potential curves for all the O_2^- states derived from unexcited O plus O^-, and one from $O(^1D)$ plus O^-. Previously, Zemke *et al.* [44] had used MCSCF methods to compute the ground-state wave function of O_2^-, giving an electron affinity for O_2 in agreement with the experimental values.

Das and Wahl [45] have used the OVC method in a study of the potential curve of the ground state of the F_2 molecule. The results (D_e, ω_e, R_e, 1.67 eV, 942 cm^{-1}, and 2.67 a.u.) agree well with experiment (1.68, 932, and 2.68, respectively). The subject is instructively reviewed in a paper by Berkowitz and Wahl on the dissociation energy of F_2 [46]. Calculations on excited states of F_2 include a paper by Demoulin and Jungen [47].

Cohen and Schneider have reported *ab initio* potential curves for the ground and singly excited states of Ne_2 [48]. Saxon and Liu have made MCSCF–CM calculations on the lowest excited triplet states $^3\Sigma_u^+$ and $^3\Sigma_g^+$ of Ar_2 and compared their results with elastic scattering data; good agreement was found for the $^3\Sigma_u^+$ curve [49]. The curve has a calculated well depth of 0.68 eV at $4.59a_0$.

Polarization–CM calculations on a number of singlet and triplet states of Zn_2 have been made by Hay *et al.* and potential curves are given [50]. Estimated curves for Hg_2 are also given.

G. MIXED V STATES

In Section III.2 (q.v.) the very special properties of the V state of H_2 have been discussed. In other molecules with more electrons than H_2, two kinds of V states, V_σ and V_π, can be predicted. In terms of MOs, V_σ states have electron configurations which include a grouping $(\sigma_a+\sigma_b)(\sigma_a-\sigma_b)$, or $\sigma\sigma^*$, and V_π states a grouping $(\pi_a+\pi_b)(\pi_a-\pi_b)$, or $\pi\pi^*$. In homopolar first-row diatomic molecules, these are $^1\Sigma_u^+$ excited states with groupings $3\sigma_g 3\sigma_u$ and $1\pi_u^3 1\pi_g$ respectively [51]. However, it appears that the *observed* V states are mixtures of V_σ and V_π with the latter predominant but the former making a rather major contribution. Such mixed V states are observed in N_2 and O_2. In F_2, Cl_2, Br_2, and I_2 there are V_σ states analogous to that of H_2. All these V states have special properties similar to those of the V_σ state of H_2.

In the case of O_2, computations on the $1\sigma_g^2 1\sigma_u^2 2\sigma_g^2 2\sigma_u^2 3\sigma_g^2 1\pi_u^3 1\pi_g^3$, $^3\Sigma_u^-$ state (B state) by Morokuma and Konishi [39], and by Schaefer and Miller [38], show a large amount of varied CM, increasing with increasing R. For the three most important CM Φ's, $\cdots 3\sigma_g^2 1\pi_u^3 1\pi_g^3$, $\ldots 3\sigma_g 1\pi_u^4 1\pi_g^2 3\sigma_u$, and $\ldots 3\sigma_g 1\pi_u^2 1\pi_g^4 3\sigma_u$, the respective coefficients according to Schaefer and Miller are 0.894, 0.367, and 0.156 at $R = 2.5$ a.u. and 0.753, 0.524, and 0.301 at $R = 3.5$ a.u. ($R_e = 3.03$ a.u.). For O_2, additional V states of symmetries $^1\Delta_u$ and $^1\Sigma_u^+$ can be predicted [51].

According to a calculation by Lefebvre-Brion, the observed V state of N_2 (the b′ state), presumably near its R_e, is a mixture of about 65% V_π and 35% V_σ. There should be another highly excited (probably unstable) state with a reverse mixture (not observed). An SCF calculation of the hypothetical *pure* V_π state of N_2 shows it to be much higher in (vertical) energy (14.03 eV above the ground state) than any of the other states ($^3\Sigma_u^+$, $^3\Delta_u$, $^3\Sigma_u^-$, $^1\Sigma_u^-$, and $^1\Delta_u$, with computed energies in the range 6–9 eV above the ground state) of the same $1\sigma_g^2 1\sigma_u^2 2\sigma_g^2 2\sigma_u^2 1\pi_u^3 3\sigma_g^2 1\pi_g$ configuration. Interaction with V_σ brings it down to 12.96 eV above the ground state but at a much larger R_e than the others; the vertical energy of this b′ state above the ground state is, however, empirically about 14.3 eV, suggesting that it is then mainly V_π. A further complication is the onset of Rydbergization at smaller R values.

Although V states have ion–pair character (H^+ plus H^-, N^+ plus N^-, O^+ plus O^-, etc.), at intermediate R values, on dissociation they actually short-cut to atom–pair states in which one atom is excited to a Rydberg state. However, in N_2 and O_2 it appears that they can short-cut to a valence-shell atom–pair state. Thus in N_2, the b′ state dissociates to s^2p^3, ^2D plus ^2P, and in O_2 the B state dissociates to s^2p^4, ^3P plus ^1D. In O_2, the B state apparently takes on increasingly V_σ character as $R \to \infty$.

H. RYDBERG STATES

Rydberg states of N_2 and O_2 are known experimentally, and have been the subject of theoretical calculations. Many of the *vibrational levels* of the Rydberg states, however, are rather strongly mixed with those of excited valence-shell states, and for comparison with theoretically computed pure

Rydberg states, a "deperturbation" procedure must first be applied [52]. This done, we can make the comparison. In other cases, however, there is extensive mixing at the electronic level (see below).

If a Rydberg state $A_2{}^+Ry$ is the lowest state of its symmetry species for a given "core" ($A_2{}^+$) state, an ordinary SCF computation can usually be made for the state. However, the basis set must include not only STFs suitable to give a good description of the core state but also extra STFs to approximate the nearly hydrogenic outer part of the Rydberg MO Ry; in practice, it is found that a *single* UAO-like STF with relatively low carefully optimized ζ ("diffuse" STF) works rather well.

For example, consider the Rydberg states $\ldots 1\pi_u{}^3 3\sigma_g{}^2 2\pi_u$, ${}^3\Sigma_g{}^+$, ${}^3\Delta_g$, ${}^3\Sigma_g{}^-$, ${}^1\Sigma_g{}^-$, ${}^1\Delta_g$, ${}^1\Sigma_g{}^+$, where the Rydberg MO $2\pi_u$ corresponds to the UAO $3p\pi$ and the core corresponds to the $\ldots 1\pi_u{}^3 3\sigma_g{}^2$, ${}^2\Pi_u$ excited state of $N_2{}^+$. All these states can be computed by the SCF method, in spite of the fact that the Rydberg configuration $\ldots 1\pi_u{}^4 3\sigma_g 4\sigma_g$ (where $4\sigma_g$ corresponds to the UAO 3s) gives a ${}^3\Sigma_g{}^+$ and a ${}^1\Sigma_g{}^+$ state of the same species as two of the above. Further, both these ${}^1\Sigma_g{}^+$ states are of the same species as the $\ldots 1\pi_u{}^4 3\sigma_g{}^2$, ${}^1\Sigma_g{}^+$ ground state, and care must be used to avoid falling into that state. With care, higher states of a Rydberg series can be obtained by SCF calculations, on readjusting the ζ of the Rydberg MO; as a guide, the empirical Rydberg series n^* values [see Eq. (III.12)] can be used.

Not many calculations on Rydberg states have been made; an early paper of Lefebvre-Brion and Moser [53] deals with N_2, although less accurately than by an SCF calculation. In a recent paper [54], we have reported accurate SCF computations of among others the several $\ldots 1\pi_u{}^3 3\sigma_g{}^2 2\pi_u$ Rydberg states mentioned in the preceding paragraph.

Since the exchange interaction between the Rydberg electron and the core is relatively small (more and more so for higher members of a Rydberg series), the necessary electron correlation by CM should be approximately just that of the core alone. In confirmation of this expectation, we find [54] that the term values (ionization energies) of the Rydberg MO for the SCF-computed $\cdots 1\pi_u{}^3 3\sigma_g{}^2 2\pi_u$ and other states agree closely with the corresponding orbital energies ε; in other words, Koopmans' theorem is rather closely observed.

Cartwright *et al.* [55] have computed the excitation energies of a series of singlet and triplet Rydberg states of O_2 with the configuration

$$\ldots 3\sigma_g{}^2 1\pi_u{}^4 1\pi_g \mathrm{Ry},$$

where Ry is one of the Rydberg MOs ns, $np\sigma$, $np\pi$, $nd\sigma$, $nd\pi$, with mostly $n = 3$–5 (or 4–6 in the united atom in the case of $np\sigma$), using "improved virtual orbitals." The results are compared with experimental evidence. In particular, the $\ldots 1\pi_g 3s$, ${}^3\Pi_g$ level is so identified. Buenker and Peyerimhoff

[56] have made all-valence-electron CM calculations for the Rydberg states for which Ry is $3p\pi(2\pi_u)$, $3s(4\sigma_g)$, or $3p\sigma$($4p\sigma$ in the united atom).

As has been noted earlier, some valence-shell states are semi-Rydberg or near-Rydberg (see Section III.B); in other words, they are Rydbergescent (Section IV.A) [57]. At sufficiently small R values, many states with promoted MOs (cf. Section II.B) become Rydberg states. How fast this happens with decreasing R varies from case to case. Some cases where already at R_e a state becomes fully a Rydberg state, and others where this does *not* happen, have been discussed in the chapter on diatomic hydrides (see Sections IV.A and IV.E).

An interesting case is that of the

$$\ldots 2\sigma_u{}^2\, 1\pi_u{}^4\, 3\sigma_g\, 1\pi_g,\ {}^3\Pi_g$$

and

$$\ldots 2\sigma_u\, 1\pi_u{}^4\, 3\sigma_g{}^2\, 1\pi_g,\ {}^3\Pi_u$$

states of N_2 [58]. At R_e, these are normal valence-shell states, where $1\pi_g$ is approximately of the form $2p\pi_a - 2p\pi_b$, but at $R = 0.9$ a.u., $1\pi_g$ has been promoted almost perfectly to a $3d\pi$ UAO, which is a fully Rydberg MO. The corresponding ${}^1\Pi_g$ and ${}^1\Pi_u$ states (not computed) may be expected to behave similarly. These various states at R_e could not be called near-Rydberg, but as R decreases they before long become *fully* Rydberg. Rydbergization of the $1\pi_g$ MO to $3d\pi$ must also occur in the several states of the configurations in $1\pi_u{}^3\, 3\sigma_g{}^2\, 1\pi_g$ at several R values, but more rapidly for the ${}^1\Sigma_u{}^+$ state than for the others.

In certain other cases having unpromoted but genuine Rydberg MOs at R_e, other types of changes occur. Thus for the essentially $3s(4\sigma_g)$ MO in the Rydberg states $\ldots 1\pi_u{}^4\, 3\sigma_g\, 4\sigma_g$ and $\ldots 1\pi_u{}^3\, 3\sigma_g{}^2\, 4\sigma_g$ at R_e the $4\sigma_g$ MO as $R \to 0$ becomes the Rydberg MO $3d\sigma$ of the united atom. (Meanwhile, the valence-shell $3\sigma_g$ MO of N_2 becomes 3s of the united atom.) On the other hand, the essentially $3p\pi(2\pi_u)$ MO in the Rydberg states $\ldots 1\pi_u{}^4\, 3\sigma_g\, 2\pi_u$ ${}^3\Pi_u$ and ${}^1\Pi_u$ at R_e becomes, shall we say, de-Rydbergized or *valenated* into $3p\pi$ of the valence shell of the united atom. These correlate as $R \to 0$ with the united-atom (Si) states $\ldots 3s3p^3$, ${}^1D$ or ${}^3D$, in which $2\pi_u$ becomes $3p\pi$ within the valence shell (although in an excited state of) the united atom. At the same time the valence-shell MOs $2\sigma_u$ and $3\sigma_g$ become the valence-shell AOs $3p\sigma$ and 3s of Si.

More generally than of Rydbergization or valenation, one may speak of *conversion*—of an MO from one form to another with changing R. For example, the valence-shell $3\sigma_g$ MO of N_2 at R_e strongly resembles $3d\sigma$ of the semi-united atom, but at smaller R is converted to almost pure 3s form, which it retains all the way to the united atom [8]. At the same time, the Rydberg MO 3s becomes converted in the united atom to a $3d\sigma$ Rydberg

MO. Rydbergization is a form of conversion of special interest. Promotion is a form of conversion that occurs on decreasing R for many valence-shell MOs, including both occupied MOs and excited MOs; in the latter case, promotion entails Rydbergization.

There are really two types of Rydbergization, which have been called MO and MO-or-state [59]. MO Rydbergization occurs notably for an antibonding promotable MO such as one which in the homopolar case has LCAO form $nx - nx$ at large R but goes to a promoted Rydberg form (higher n) as $R \to 0$. In the process, valenation may or may not occur. At intermediate R values, the form of such an MO can be represented conveniently as a linear combination of a valence shell $(nx - nx)$ and a Rydberg function, but it is a misconception (though frequently entertained) to think of it as a mixture of two entities capable of independent existence. It is characteristic of the MO that it changes its character with changing R. The simplest examples are the MOs of H_2^+. In MO Rydbergization, one also cannot correctly say that there is mixing of a Rydberg *state* and a valence-shell state (further, see below). The MO and the state are simply of intermediate type.

As an example of MO-or-state Rydbergization, the 4σ MO in the $1\sigma^2 2\sigma^2 3\sigma 4\sigma$, ${}^3\Sigma^+$ repulsion state of BH (see Section IV.A) has the antibonding LCAO form $a1s_H - b2p\sigma_B$ at large R, but at R_e has become Rydbergized to a 3s MO which remains a Rydberg AO in the united-atom carbon. Instead of thinking in terms of Rydbergization of the 4σ MO, an alternative point of view is applicable in this and similar cases; namely, one can say that the ${}^3\Sigma^+$ *state*, which at large R is a valence-shell state, is transformed adiabatically to a Rydberg state at small R as a result of an avoided crossing between a diabatic Rydberg-type ${}^3\Sigma^+$ potential curve dissociating to the Rydberg atomic state $1s^2 2s^2 3s$, ${}^2S$ of boron (plus 1s of hydrogen) and a diabatic valence-shell repulsive ${}^3\Sigma^+$ curve rising rapidly with decreasing R from a $1s^2 2s^2 2p$, ${}^2P$ boron (plus 1s hydrogen) asymptote. From this *state* point of view, it is readily understandable that, close above the ${}^3\Sigma^+$ Rydberg state curve just discussed, there exists a corresponding ${}^1\Sigma^+$ Rydberg curve with a minimum at about the same R_e as the ${}^3\Sigma^+$ curve, but whose dissociation asymptote is necessarily the Rydberg atomic state $1s^2 2s^2 3s$, ${}^2S$ (plus 1s hydrogen).

The case of MO Rydbergization is very different in that no alternative viewpoint of mixing of diabatic Rydberg and valence-shell states is admissible.

Somewhat similar to the BH case is a situation in O_2, where BP (Buenker and Peyerimhoff) [60] have made *ab initio* calculations on the mixing of valence-shell and Rydberg states. The X ${}^3\Sigma_g^-$ ground state of O_2 ($R_e =$ 2.28 a.u.) belongs to the electron configuration $\ldots 3\sigma_g^2 1\pi_g^4 1\pi_g^2$, and on dissociation goes to two $\ldots 2p^4$, ${}^3P$ ground-state atoms. The best-known excited state is the B $^3\Sigma_u^-$ state, a mixed V state, belonging to the configuration $\ldots 1\pi_u^3 1\pi_g^3$ predominantly, mixed with some $\ldots 3\sigma_g 1\pi_g^4 1\pi_g^2 3\sigma_u$. Predicted to overlap the intense B$\leftarrow$X spectroscopic transition is a weaker transition $\ldots 3\sigma_g^2 1\pi_u^4 1\pi_g 3\sigma_u$, $^3\Pi_u \leftarrow$ X whose $^3\Pi_u$ upper state is held responsible for some of the predissociation observed in the B state.

For the $^3\Pi_u$ state just mentioned, BP have made a CM (configuration mixing) calculation from which they conclude that the $3\sigma_u$ MO at large R values has the valence-shell LCAO form $2p\sigma - 2p\sigma$,[7] but at small R values has a Rydberg form which they call $3p\sigma$ but which in the united atom would be $4p\sigma$. As they point out, this MO transformation (similar to that already noted in BH) is equivalent to the occurrence of CM and an avoided crossing between a diabatic pure Rydberg and a diabatic pure valence-shell $^3\Pi_u$ curve, resulting in two adiabatic curves. The two diabatic curves are what one would deal with in an SCF calculation. The lower of the two adiabatic curves has an asymptote as $R \to \infty$ of two ^{3}P normal O atoms; with decreasing R it rises steadily, then dips according to the calculation to a small Rydberg minimum at about 2.18 a.u. The upper adiabatic curve, with a minimum at about 2.45 a.u., must dissociate to yield a Rydberg atomic state of configuration $\ldots 2p^3 3s$ plus a normal O atom. In the lower adiabatic curve, $3\sigma_u$ changes from ($4p\sigma$, $3s-3s$) character at small R to ($4f\sigma$, $2p\sigma - 2p\sigma$) character at large R. Here the first symbol within the parentheses is the united-atom form of the MO, and the second is the large-R LCAO form. In the upper curve, the excited MO would now be $4\sigma_u$ (or perhaps $5\sigma_u$ or higher as $R \to 0$), of form ($4f\sigma$, $2p\sigma - 2p\sigma$) at small R and form ($4p\sigma$, $3s-3s$) at large R. Note that $4f\sigma$ and $4p\sigma$ are the promoted united-atom forms to which $2p\sigma - 2p\sigma$ and $3s-3s$ respectively would go on the basis of MO Rydbergization. These correlations can be seen most simply by referring to the MO correlation diagrams of H_2^+ and H_2. In passing, we note that besides the two $^3\Pi_u$ states just discussed, there should be two corresponding $^1\Pi_u$ states, with a diabatic Rydberg and a diabatic valence-shell state interacting to give adiabatic states. The case here is in sharp contrast to that in BH, where interaction and mixing occur for a triplet state only.

A situation analogous to that in the $^3\Pi_u$ and $^1\Pi_u$ states of O_2 just discussed is predicted to exist for the experimentally well-known $^3\Sigma_u^+$ and $^1\Sigma_u^+$ Rydberg states of N_2 which at normal R values have the electron configuration $\ldots 1\pi_u^4 3\sigma_g 3\sigma_u$, with $3\sigma_u$ a Rydberg MO of the form $4p\sigma$, $3s-3s$. At large R values, there must exist a $^3\Sigma_u^+$ and a $^1\Sigma_u^+$ state of the same formal configuration $\ldots 1\pi_u^4 3\sigma_g 3\sigma_u$, but with $3\sigma_u$ now a valence-shell MO of form $4f\sigma$, $2p\sigma - 2p\sigma$. The latter states must have rapidly rising repulsive potential curves. A little examination shows that the diabatic curves of the small-R Rydberg and the large-R valence-shell states, unlike the Π_u curves in O_2,

[7] Here the z axes of the two centers are assumed directed toward each other.

would intersect only at very large energies, so that it is rather useless to look for the corresponding adiabatic curves. However, one can say that in the *lower* of the adiabatic curves (for $^3\Sigma_u^+$ or for $^1\Sigma_u^+$) the $3\sigma_u$ MO would go from the 4pσ, 3s−3s form at small R to the 4fσ, 2pσ−2pσ form at large R. Actually, the $^1\Sigma_u^+$ Rydberg curve soon crosses *with only small interaction* [61] another $^1\Sigma_u^+$ valence-shell (or semi-Rydberg) curve, the mixed-V b′ curve (mixed $\ldots 1\pi_u^3 3\sigma_g 1\pi_g$ and $\ldots 1\pi_u^4 3\sigma_g 3\sigma_u$, probably mostly the former near the crossing point, but a 50–50 mixture as $R \to \infty$ according to recent calculations [62]). But that is another story.

A situation similar to that for the $^3\Sigma_u^+$ and $^1\Sigma_u^+$ states just discussed exists for the $\ldots 1\pi_u^3 3\sigma_g^2 3\sigma_u$, $^3\Pi_g$ and $^1\Pi_g$ states, of which the latter is experimentally well known. Here again, the diabatic Rydberg states would dissociate to a $\ldots 2s^2 2p^2 3s$, ^4P Rydberg atom plus a normal atom. On the other hand, there is according to Michels [63] a repulsive $^1\Pi_g$ state that, along with the stable valence-shell state $\ldots 1\pi_u^4 3\sigma_g 1\pi_g$, $^1\Pi_g$, dissociates to two $\ldots s^2 p^3$, ^2D atoms. This repulsive state at large R values must have the electron configuration $\ldots 1\pi_u^3 3\sigma_g^2 3\sigma_u$ like the Rydberg state at small R values. If there is an avoided crossing of the potential curves of the two diabatic states in the present case, it must occur at very high energy values. Similar relations hold for the $^3\Pi_g$ curves.

Another example of MO-or-state Rydbergization more like that in O_2 is found in NO [64]. Here the diabatic curve of the well-known D, $^2\Sigma^+$ Rydberg state of configuration $\ldots 1\pi^4 6\sigma$, where 6σ is 4pσ, 3s−3s, is predicted to interact on increasing R with the repulsive valence-shell diabatic A′ curve of configuration $\ldots 1\pi^4 6\sigma$ with 6σ now of the form 4fσ, 2pσ−2pσ. Here, after dealing with some additional complications, Gallusser and Dressler using perturbation methods have derived the form of the lower adiabatic curve from the interaction.

So much for the two types of Rydbergization. BP have also discussed a different type of avoided crossing in O_2, between a pure Rydberg and a pure valence *state* of different formal configuration [56, 65]. Here the pure valence diabatic state is the $1\pi_u^3 1\pi_g^3$, $^3\Sigma_u^-$ B state, while the pure Rydberg diabatic state is the $1\pi_u^4 1\pi_g 2\pi_u$, $^3\Sigma_u^-$ state, where $2\pi_u$ is (3pπ, 3pπ+3pπ). Experimentally the B state, which is the lower of the two resulting adiabatic states, is well known at fairly large R values where it is nearly pure valence state, but the calculations show that because of mixing of the two configurations mentioned, Rydberg character must begin to be predominant even at R_e of the ground state. It is interesting also to consider what must happen as $R \to 0$. For the united-atom sulfur the electron configuration becomes $1s^2 2s^2 2p\sigma^2 2p\pi^3 3p\pi 3d\pi$, in which $1\pi_g$ is Rydbergized to 3dπ but $2\pi_u$ is valenated to 3pπ of the valence shell of the united atom.

Of further interest, the upper of the two adiabatic states resulting from the interaction just discussed has been identified with an experimentally known excited state of higher energy with an unusually high vibration frequency [65, 66].

The same electron Rydberg and valence-shell configurations that give rise to the B state also give rise to other states. Among these are a $^3\Delta_u$ Rydberg and a $^3\Delta_u$ valence-shell state, whose interaction BP have also computed and show to be small. This result shows that the occurrence or nonoccurrence of strong interaction cannot be predicted from electron configurations alone, but depends in a nonobvious way on the magnitudes of matrix elements.

The preceding discussion has led to the identification of two limiting types of interaction between pure-electron-configuration diabatic states. (Of course it must always be kept in mind that actual states always involve CM, although in most cases there is one predominant configuration which one may hypothesize to exist in pure form.) Among other possibilities, one of the two states may be a Rydberg state and the other a valence-shell state. In brief, two diabatic states of the same symmetry species but different electron configurations may interact

(a) only at the gyrovibronic level; or
(b) already at the electronic state level.

In case (a), the interaction occurs only between individual vibration–rotation levels of the two states. Many examples of interaction at this level between Rydberg and valence-shell states are known, especially in N_2 [61] and NO. The Δ_u states of O_2 mentioned previously must also come under this case. In case (b), if the potential curves of the two diabatic states cross, they may undergo a more or less strongly avoided crossing to give two new adiabatic states when CM is considered. Whether case (a) or case (b) is realized in a particular example depends on the magnitude of interaction matrix elements.

In MO Rydbergization, we do *not* have an example of case (b) interaction. Instead, there is a single diabatic state in which the form of a single MO changes with R, from valence-shell form for large R, through intermediate forms, to Rydberg forms at small R. But there are no separate valence-shell and Rydberg state curves which interact. On the other hand, in MO-or-state Rydbergization, although the same formulation in terms of the changing form of a single MO with R is valid, an alternative formulation in terms of the interaction of two separate diabatic curves as in case (b) is *also* valid.

To further illustrate MO Rydbergization, a very simple example is $1\sigma_u$ of H_2, of form $1s-1s$ at large R and form $2p\sigma$ as $R \to 0$. Note that at small R, $1\sigma_u$ becomes a true Rydberg MO, as in the V state $1\sigma_g 1\sigma_u$, $^1\Sigma_u{}^+$. Although at intermediate R values $1\sigma_u$ has the form $a(1s-1s)+b\ 2p\sigma$, suggesting perhaps that it is a mix of two independent MOs $1s-1s$ and $2p\sigma$ and the V state a mix of a Rydberg state $1\sigma_g 2p\sigma$, ${}^1\Sigma_u{}^+$ and a valence-shell state $1\sigma_g(1s-1s)$, ${}^1\Sigma_u{}^+$, this beguiling idea is not justified. There is no independent Rydberg state $1\sigma_g 2p\sigma$, ${}^1\Sigma_u{}^+$. If one is skeptical, note that the V state (at moderately small R values corresponding to its R_e where it is fairly well Rydbergized) fits smoothly into a Rydberg series $1\sigma_g np\sigma_u$, ${}^1\Sigma_u{}^+$, where $np\sigma$ at large R is of the LCAO form $np\sigma - np\sigma$. That it fits smoothly into a Rydberg series can be seen from a consideration of the quantum defects of successive members of the series: $\delta = 0.210$ for $2p\sigma$ in the V state, $\delta = 0.196$ for $3p\sigma$ in $1\sigma_g 3p\sigma$, ${}^1\Sigma_u{}^+$, and $\delta = 0.187$ for $4p\sigma$ in $5p\sigma$, ${}^1\Sigma_u{}^+$. A similar discussion applies in other cases, for example to the $1\pi_g$ MO in H_2 or N_2 which is $2p\pi - 2p\pi$ as $R \to \infty$ but soon becomes $3d\pi$ as $R \to 0$.

I. BONDING AND BINDING

1. Uncorrelated Wave Functions

In a molecule whose atoms are far apart, there is no net force between the atoms. However, this zero force is the sum of repulsions between the nuclei and attractions between each nucleus and the electrons of the other atoms. The zero net force is a result of the cancellation or *shielding* of the repulsive by the attractive forces. Now if the atoms approach each other, the charge distributions of the electrons are distorted to give a net force of attraction or repulsion. According to the Hellmann–Feynman theorem [67], the attractive force exerted by the electrons on a nucleus of charge Ze can be computed correctly by assuming that each element dq of electronic charge in the quantum mechanically computed charge distribution exerts a classical force $e^2 Z\, dq/r^2$ directed toward the nucleus, where r is the distance of the electron from the nucleus.

Thus in a diatomic molecule AB, the resultant force (in atomic units) on nucleus A, a force which by symmetry is directed along the line joining the two nuclei, is

$$F_A = Z_A Z_B/R^2 - Z_A \int [\rho(r_A, \theta_A) \cos\theta_A / r_A{}^2]\, dV, \tag{2}$$

where ρ is the electronic charge density dq/dV (independent of ϕ) at a location r_A, θ_A as measured from nucleus A, with $\theta_A = 0$ on the axis in the direction of B. A similar expression holds for nucleus B. Further, $F_A = F_B = -dU/dR$, if attractive forces are taken as positive.

In a stable molecule at R_e, $F_A = F_B = 0$; the electronic attraction forces, or *binding forces*, just balance the internuclear repulsion. This result is exact if ρ is calculated from the exact wave function. It is also exact for the ρ computed from an SCF wave function at the *computed* R_e.

As Berlin [68] first pointed out, the charge distribution around the nuclei in a stable molecule at equilibrium can be divided into *binding regions*, where the electronic forces pull the nuclei together, and *antibinding regions*, where they push them apart. For the ground state of H_2^+, the boundaries of these two regions are shown by dashed lines in the upper right hand part of Fig. II.3. The net effect of the electron or electrons in the binding and antibinding regions is at R_e an attractive force just equal to the nuclear repulsion. In Fig. 3 of this chapter the boundaries between binding and antibinding regions in the first-row homonuclear diatomic molecules at R_e are indicated by dotted-line curves.

In an interesting series of papers [19, 69], Bader and collaborators have analyzed in detail the role of the Hellmann–Feynman forces in chemical binding. First, they define dimensionless quantities f such that in a molecule AB the total force on nucleus A is

$$F_A(R) = (Z_A/R^2)(Z_B - f_A), \tag{3}$$

where

$$f_A(R) = R^2 \int [\rho(\theta_A, r_A) \cos\theta_A / r_A{}^2] \, dV.$$

Then f_A is R^2 times the force exerted by the electrons on unit nuclear charge A, in other words, R^2 times the field of the electrons at A. In SCF approximation, f_A can be written as a sum over the effects f_{iA} of the electrons in individual MOs ϕ_i:

$$f_A = \sum_i f_{iA} = R^2 \sum_i N_i \left[\int \phi_i{}^* \cos\theta_A \, \phi_i / r_A{}^2 \right] d\tau, \tag{4}$$

where N_i is the number of electrons in ϕ_i. In homonuclear molecules, $f_B = f_A = f$, say, and $f_{iB} = f_{iA} = f_i$. For the first-row homonuclear molecules at R_e, Bader *et al.* [19] give a table of f_i values, which we have reproduced in Table 18. Note that the total f value $(\sum_i f_i)$ when multiplied by $Z_A/R_e{}^2$ and subtracted from $Z_A Z_B/R_e{}^2$ gives in the last column a nearly zero net force (except for Be_2, which is unstable).

In Table 18, it may seem surprising that $f(1\sigma_g)$ and $f(1\sigma_u)$ can depart so much from 1 in Li_2. The reason can be seen in the form of f_{iA} in Eq. (4). In this molecule the 1σ MO is rather strongly polarized *away* from the center of the molecule as a result of the accumulation of negative charge in the $2\sigma_g$ MO in the middle of the molecule (see Fig. 3 or 4); this polarization results in a relatively large repulsive contribution to f_A because it occurs in a region of small r_A values; the effect is then greatly amplified by the presence of the $R_e{}^2$ factor in the equation for f_{iA} ($R \gg \bar{r}_A$). Similar but smaller effects

TABLE 18

Orbital Forces in Homonuclear Diatomic Molecules

Ground state	Molecule	$f(1\sigma_g)$	$f(1\sigma_u)$	$f(2\sigma_g)$	$f(2\sigma_u)$	$f(1\pi_u)$	$f(3\sigma_g)$	$f(1\pi_g)$	$\sum_i f_i$	Net force (a.u.)
$^1\Sigma_g^+$	Li_2	0.706	0.658	1.591					2.955	0.005
$^1\Sigma_g^+$	Be_2 [a]	1.051	1.028	2.003	−0.399				3.683	0.103
$^3\Sigma_g^-$	B_2	0.979	0.971	2.305	−0.492	1.188			4.951	0.027
$^1\Sigma_g^+$	C_2	0.969	0.954	2.250	−0.436	1.125 [b]			5.987	0.015
$^1\Sigma_g^+$	N_2	1.160	1.085	2.682	−0.463	1.216 [b]	0.150		7.046	−0.075
$^3\Sigma_g^-$	O_2	1.232	1.138	2.934	−0.518	1.302 [b]	0.174	0.426	7.990	0.016
$^1\Sigma_g^+$	F_2	1.243	1.123	2.447	−0.168	1.232 [b]	0.516	0.656 [b]	8.937	0.080

[a] The $^1\Sigma_g^+$ state of Be_2 is repulsive. These results refer to $R = 3.5$ bohrs.

[b] All f_i values for comparative purposes are quoted for double occupation of the orbitals. Values marked by [b] are to be doubled in obtaining the total electronic force, since they refer to π orbitals containing four electrons.

occur in $1\sigma_g$ and $1\sigma_u$ in the other molecules, as a result of polarization of these, in some cases away from the center, in other cases toward the center.

Further [19], f_{iA} can be broken up as[8]

$$f_{iA} = f_{iA}^{AA} + f_{iA}^{AB} + f_{iA}^{BB}. \tag{5}$$

A complementary expression holds for f_{iB}. The "atomic force" f_{iA}^{AA}/R^2 is the force exerted on each unit of charge of nucleus A by the *net atomic population* (as defined in Section II.F) on A in ϕ_i; with $N_i = 2$, this population for a minimal basis set is $1/(1+S)$ electrons. At this point it should be stressed that, like the population analysis itself, this breakdown, while qualitatively instructive, is dependent on the basis set and is open to question as to quantitative meaning—see Section II.F and the discussion at the end of Section VI.C; however, the overall f_i and f values depend only on the charge distributions, which are, for the most part, independent of the basis sets. The "*overlap force*" f_{iA}^{AB}/R^2 is the force exerted on unit charge of either nucleus by the *overlap population* (see Section II.F); for $N_i = 2$, this population is $S/(1+S)$. The "screening force" f_{iA}^{BB}/R^2 is the force on unit charge of A exerted by the *net atomic* $1/(1+S)$ *population* on B; it is a measure of the electronic shielding of unit charge on nucleus B from nucleus A by electrons belonging to B only.

Table 19 gives SCF values for all the MOs of O_2 at R_e. Except for $1\sigma_g$

TABLE 19

Force Coefficients for the Orbital Densities of the Ground State of O_2 at Its Experimental R_e[a,b]

	f_{iA}^{AA}	f_{iA}^{AB}	f_{iA}^{BB}	f_i
$1\sigma_g$	0.231 (0)	0.000 (0)	1.001 (1)	1.232 (1)
$1\sigma_u$	0.129 (0)	0.010 (0)	0.999 (1)	1.138 (1)
$2\sigma_g$	0.404 (0)	1.628 (0)	0.902 (1)	2.934 (1)
$2\sigma_u$	−1.048 (0)	−0.267 (0)	0.797 (1)	−0.519 (1)
$3\sigma_g$	−2.334 (0)	1.693 (0)	0.815 (1)	0.174 (1)
$1\pi_u$	0.434 (0)	0.842 (0)	1.328 (1)	2.604 (2)
$1\pi_g$	−0.100 (0)	−0.419 (0)	0.945 (1)	0.426 (1)
Totals (f):	−2.284 (0)	3.487 (0)	6.788 (8)	7.990 (8)

[a] From Fig. 6 of Bader *et al.* [19].

[b] The values for the separated atoms ($R \to \infty$) are given in parentheses.

[c] Note that the f_{iA}^{AA}, f_{iA}^{AB}, and f_{iA}^{BB} values and their sums, but not the f_i, are dependent on the basis set used.

[8] They write, apparently erroneously, $f_i = R^2(f_i^{AA} + f_i^{BB} + f_i^{AB})$.

and $1\sigma_u$, all the σ f_{iA}^{BB} fall considerably below their $R \to \infty$ value of 1. Because of polarization, the f_{iA}^{AA} deviate considerably from zero, either positively or (strongly in $2\sigma_u$ and $3\sigma_u$) negatively. The f_{iA}^{AB} are positive except in $2\sigma_u$ and $1\pi_g$, and largest in the strongly bonding $2\sigma_g$ MO.

A breakdown into f_{iA}^{AA}, f_{iA}^{AB}, and f_{iA}^{BB} values for B_2, C_2, O_2, N_2, Li_2 and F_2 is given in Table 20. The dissociation energies D_e are seen to strongly parallel the overlap contributions to the force coefficients. It is seen also that increased D values parallel the screening deficits, that is, the deviations of the f_{iA}^{AA} and f_{iA}^{AB} values from their value as $R \to \infty$. These relations correspond to the fact that during bonding, charge is withdrawn from the atoms and transferred into the overlap region.

TABLE 20

Force Coefficients for the Orbital Densities of Ground States at R_e [a, b]

	f_{iA}^{AA}	f_{iA}^{AB}	f_{iA}^{BB}	f	D_e (eV)
Li_2	−0.563 (0)	0.927 (0)	2.591 (3)	2.955 (3)	1.106
B_2	−0.644 (0)	1.708 (0)	3.887 (5)	4.951 (5)	2.884
C_2	−0.735 (0)	2.198 (0)	4.523 (6)	5.987 (6)	6.251
N_2	−1.943 (0)	3.853 (0)	5.136 (7)	7.046 (7)	9.909
O_2	−2.284 (0)	3.486 (0)	6.788 (8)	7.990 (8)	5.181
F_2	−1.949 (0)	2.505 (0)	8.381 (9)	8.937 (9)	1.647

[a] From Bader *et al.* [19], Table VII.
[b] The values in parentheses are for the separated atoms ($R \to \infty$).
[c] Note that the individual values but not the total f are dependent on the basis set used.

MOs are defined as *binding* or *antibinding* according as f_{iA} ($=f_{iB}$ in homopolar molecules as here) at R_e is greater or less than its value as $R \to \infty$, *and nonbinding* if f_i is near its value for large R. For the most part, bonding MOs are binding, antibonding MOs antibinding, and nonbonding MOs are nonbinding [19]. Exceptions are $3\sigma_g$ and $1\pi_g$. Further interesting discussion on this question, and on what happens to R_e on ionization of an electron from each of the various types of MOs is given in Ref. 19. In general, the active role of the Hellmann–Feynman forces in determining R_e is evident. Likewise for repulsion states, as when two He atoms approach, this interaction correlates with negative values of the Hellmann–Feynman forces; and, more in detail, one can see how the repulsive force of the $1\sigma_u$ electrons in the $1\sigma_g{}^2 1\sigma_u{}^2$ electron configuration predominates over the attractive force of the $1\sigma_g$ electrons [69].

2. Correlated Wave Functions

The discussion in Section 1 has been applied there only to SCF wave functions, nearly all at R_e of stable molecules, where the Moeller–Plesset theorem holds that the charge densities should be correct to second order (cf. Section IV.D). The theorem is relevant when, as in most cases near R_e, there is just one predominant electron configuration [cf. Eq. (IV.2)] in the wave function including electron correlation. However, in most ground states as $R \to \infty$, one or more additional configurations become of major importance, for example, $1\sigma_u{}^2$ becomes as important as $1\sigma_g{}^2$ as $R \to \infty$ in H_2 (cf. Chapter III). Here a breakdown of Hellmann–Feynman forces in terms of MOs cannot be made.

Bader and Chandra [69b], using OVC functions of Das and Wahl (see Section III.D), have made density difference contour maps and studied f values as a function of R for H_2. In analogy to Eq. (5), f_A can be broken down into three terms:

$$f_A = f_A^{AA} + f_A^{AB} + f_A^{BB}. \tag{6}$$

Here the atomic densities of Eq. (3) are computed from the (not quite completely) correlated OVC wave functions, and from them difference density maps (relative to superposed densities of the two separate atoms) are computed at various R values. The results for H_2, reproduced in Fig. 11, are instructive, as are the graphs of the terms of Eq. (5) shown in Fig. 12.

Consider the changes as the separated atoms approach each other. At large R (see $R = 8.0$ a.u. in Fig. 11) there is no overlap of $1s_A$ and $1s_B$, but the London dispersion forces cause a small polarization of each atom toward the other (cf. Section III.B and especially Section III.H). The polarization of atom A creates a considerable f_A^{AA}, although the corresponding force $R^2 f_A^{AA}$ is very small, while the polarization of B creates a small increment to the shielding force f_A^{BB} (see Fig. 12). Then by symmetry, $f_B = f_A$. At smaller R, overlap of $1s_A$ and $1s_B$ begins, and increases with decreasing R (Fig. 11). Electronic charge increases between the nuclei and is depleted behind the nuclei. The enlarging region of increased charge density finally spreads part way behind the nuclei (Fig. 11); corresponding changes occur in f_A and its parts (Fig. 12).

Bader and Chandra have also discussed Li_2 and He_2 in a similar way. Li_2 differs very considerably from H_2 in some interesting respects. In H_2 and Li_2, f_A is always positive, but in He_2 (except in the dispersion-force region at large R), it is negative. In He_2, f_A can be broken down into a positive f_{iA} for $1\sigma_g{}^2$ and a negative one for $1\sigma_u{}^2$, with the latter larger in magnitude than the former. (Note that in He_2, the SCF wave function is good all the way to dissociation, so that the single electron configuration $1\sigma_g{}^2 1\sigma_u{}^2$ gives an excellent approximation, except at small R.)

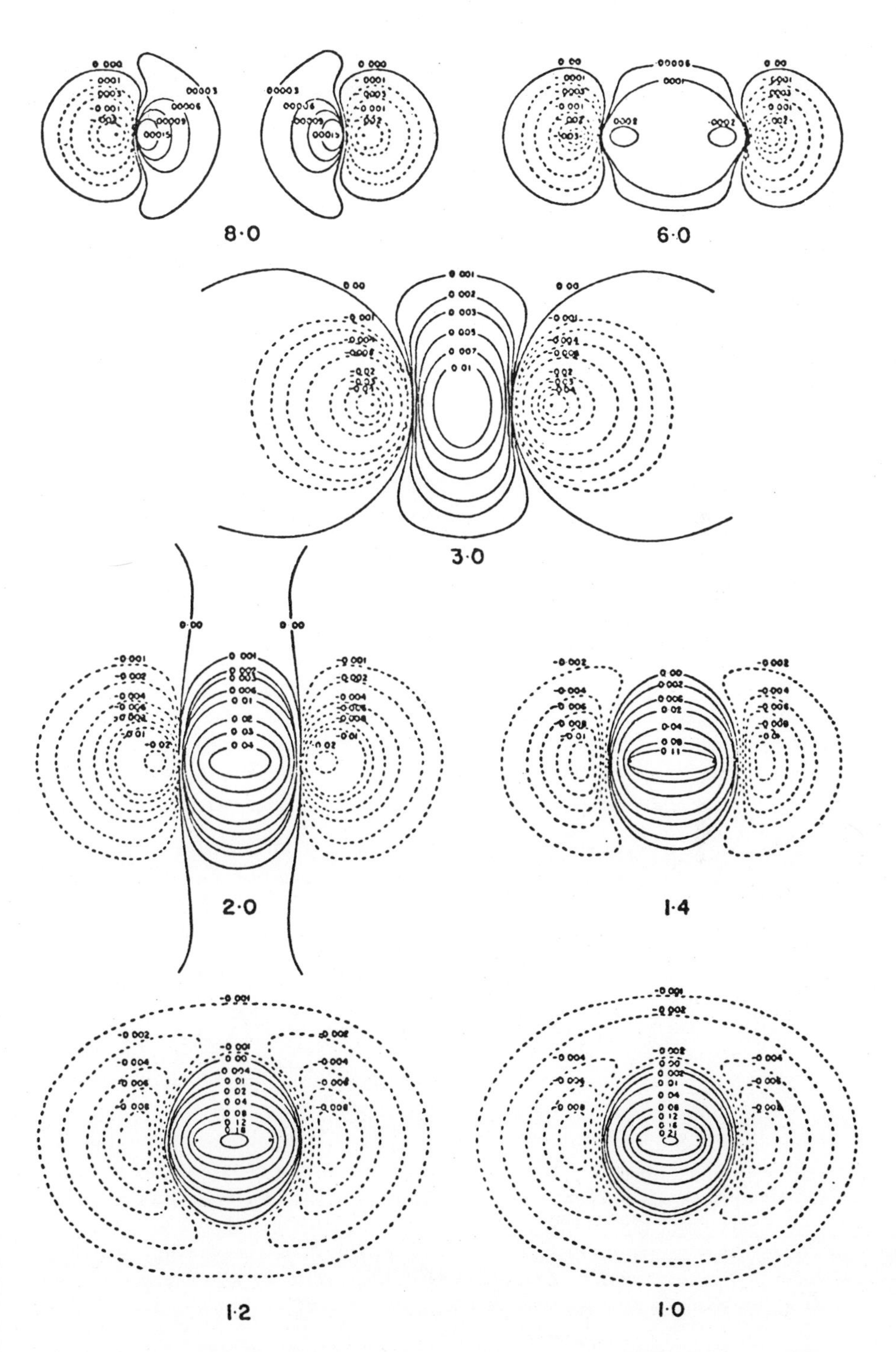

FIG. 11 Density difference contour maps for H_2 (in a.u.) at the indicated internuclear distances. [From R. F. W. Bader and A. K. Chandra. Reproduced by permission of the National Research Council of Canada from the *Canadian Journal of Chemistry*, Volume 46, pp. 953–966, 1968.]

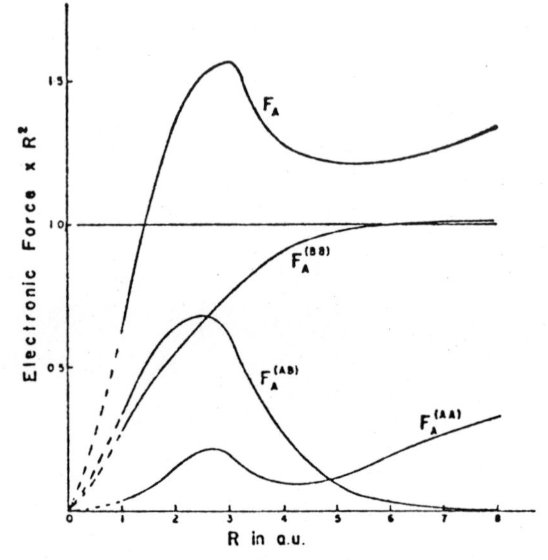

FIG. 12 The variation with R of the charge equivalents of the electronic force and its atomic, overlap, and screening components. [From R. F. W. Bader and A. K. Chandra. Reproduced by permission of the National Research Council of Canada from the *Canadian Journal of Chemistry*, Volume 46, pp. 953–966, 1968.]

J. THE COMPUTATION OF MISCELLANEOUS PROPERTIES

Given a good wave function for a molecule, numerous properties can be computed relatively easily. Some such computations have been made routinely when wave functions were obtained, but very often not. For heteropolar molecules, dipole moments, and often their variation with R, are usually computed (for examples, cf. Chapters IV and VI). Quadrupole moments are often computed. The polarizability of H_2 has been considered in Section III.G. Potential curves are very frequently computed, as mentioned in several chapters (in particular, cf. Section III.C). Spectroscopic transition probabilities (cf. Sections II.D and III.F) are frequently computed.

A sampling from recent papers of some specific computations is: quadrupole moment of CO, N_2, and NO^+ [70]; some polarizabilities [71]; effective spin–spin constants of $^3\Sigma$ excited states of N_2 and CO [72]; magnetic properties of AlH and N_2 [73]; zero-field splitting in the $b^3\,\Sigma_g^-$ state of C_2 [74].

REFERENCES

1. (a) C. W. Scherr, *J. Chem. Phys.* **23**, 569 (1955).
   (b) A. Recknagel, Z. *Phys.* **87**, 375 (1934).
2. P. E. Cade and A. C. Wahl, *Atomic Data* **13**, 339 (1974).

3. R. C. Raffenetti, *J. Chem. Phys.* **59**, 5936 (1973).
4. T. L. Gilbert and A. C. Wahl, *J. Chem. Phys.* **55**, 5247 (1971).
5. G. A. Henerson, W. T. Zemke, and A. C. Wahl, *J. Chem. Phys.* **58**, 2654 (1973).
6. W. Müller and M. Jüngen, *Chem. Phys. Lett.* **40**, 199 (1976).
7. R. S. Mulliken, *Rev. Mod. Phys.* **14**, 40 (1932).
8. R. S. Mulliken, *Chem. Phys. Lett.* **14**, 137 (1972); *Int. J. Quantum Chem.* **8**, 817 (1974).
9. J. S. Briggs and M. R. Hayns, *J. Phys. B.* **6**, 514 (1973).
10. W. C. Ermler, R. S. Mulliken, and A. C. Wahl, *J. Chem. Phys.* **66**, 3031 (1977).
11. W. A. Bingel, *J. Chem. Phys.* **30**, 1250 (1959).
12. R. S. Mulliken, *in* "Quantum Theory of Atoms, Molecules, and Solids" (P. O. Löwdin, ed.), p. 231. Academic Press, New York, 1966.
13 (a) P. S. Bagus and H. F. Schaefer, III, *J. Chem. Phys.* **56**, 224 (1972).
(b) P. S. Bagus, M. Schrenck, D. W. Davis, and D. A. Shirley, *Phys. Rev.* **9A**, 1090 (1974).
(c) D. W. Davis, R. L. Martin, M. S. Banra, and D. A. Shirley, *J. Chem. Phys.* **49**, 4235 (1973).
14. See H. T. Jonkman, Thesis, Groningen, 1975.
15. (a) C. A. Coulson and W. E. Duncanson, *Proc. Cambridge Phil. Soc.* **37**, 55, 67, 74, 397, 406 (1941); **38**, 100 (1942); **39**, 180 (1943); *Proc. Phys. Soc. London* **57**, 190 (1945); **60**, 175 (1948).
(b) I. R. Epstein, *in* "International Review of Science: Theoretical Chemistry: Physical Chemistry" (A. D. Buckingham and C. A. Coulson, eds.), Ser. 2, Vol. 1, pp. 107–161. Butterworth, London, 1975.
16. P. E. Cade and W. H. Henneker, to be published.
17. P. Eisenberger, W. H. Henneker, and P. E. Cade, *J. Chem. Phys.* **56**, 1207 (1972).
18. R. A. Tawil and S. R. Langhoff, *J. Chem. Phys.* **63**, 1572 (1975).
19. R. F. W. Bader, W. H. Henneker, and P. E. Cade, *J. Chem. Phys.* **46**, 3341 (1967).
20. A. C. Wahl, *Science* **151**, 961 (1966).
21. D. H. Boyd, *J. Chem. Phys.* **52**, 4846 (1970), Figs. 10–12.
22. D. A. Kohl and L. S. Bartell, *J. Chem. Phys.* **51**, 2896 (1969).
23. S. Huzinga, *Mem. Fac. Sci. Kyushu Univ.* **3**, 57 (1963); R. S. Mulliken, *Int. J. Quantum Chem.* **1**, 103 (1967); C. Hollister and O. Sinanoglŭ, *J. Amer. Chem. Soc.* **88**, 13 (1966).
24. R. S. Mulliken, *J. Chem. Phys.* **19**, 912 (1951); **23**, 2338 (1955).
25. R. S. Mulliken and B. Liu, *J. Am. Chem. Soc.* **93**, 6738 (1971).
26. W. G. Richards, T. E. H. Walker, and R. K. Hinkley, "A Bibliography of *Ab Initio* Molecular Wave Functions," Oxford Univ. (Clarendon) Press, London and New York, 1971; W. G. Richards, T. E. H. Walker, L. Farnell, and P. R. Scott, "Supplement for 1970–1973," Oxford Univ. (Clarendon) Press, London and New York.
27. On Li_2: G. Das and A. Wahl, *J. Chem. Phys.* **44**, 87 (1966); on Li_2: G. Das, *ibid.* **46**, 1568 (1967); on F_2: G. Das and A. Wahl, *Phys. Rev. Lett.* **24**, 440 (1970).
28. R. J. Buenker and S. D. Peyerimhoff, *Theor. Chim. Acta* **35**, 33 (1974), and personal communication.
29. H. F. Schaeffer, III, and F. E. Harris, *J. Chem. Phys.* **48**, 4946 (1968).
30. H. H. Michels, *J. Chem. Phys.* **53**, 841 (1970).
31. E. W. Thulstrup and A. Andersen, *J. Phys. B.* **8**, 965 (1975).
32. D. C. Cartwright and T. H. Dunning, Jr., *J. Phys. B.* **8**, L100 (1975).
33. P. F. Fougere and R. K. Nesbet, *J. Chem. Phys.* **44**, 285 (1966).
34. C. F. Bender and E. R. Davidson, *J. Chem. Phys.* **47**, 4972 (1967).
35. (a) C. F. Bender and E. R. Davidson, *J. Chem. Phys.* **46**, 3313 (1967).
(b) M. Dupuis and B. Liu, *J. Chem. Phys.*, to be published.
36. S. R. Langhoff and E. R. Davidson, *Int. J. Quantum Chem.* **8**, 61 (1974).

37. H. F. Schaefer, III, *J. Chem. Phys.* **54**, 2207 (1971).
38. H. F. Schaefer, III, and W. H. Miller, *J. Chem. Phys.* **55**, 4107 (1971).
39. K. Morokuma and H. Konishi, *J. Chem. Phys.* **55**, 402 (1971).
40. S. D. Peyerimhoff and R. J. Buenker, *Chem. Phys. Lett.* **16**, 235 (1972); *Chem. Phys.* **8**, 324 (1975); P. J. Hay, *J. Chem. Phys.* **59**, 2468 (1973).
41. B. J. Moss and W. A. Goddard, III, *J. Chem. Phys.* **63**, 3523 (1975); B. J. Moss, F. W. Bobrowicz, and W. A. Goddard, III, *ibid.* **63**, 4632 (1975).
42. P. S. Julienne, D. Nuemann, and M. Krauss, *J. Chem. Phys.* **64**, 2990 (1976).
43. M. Krauss, D. Nuemann, A. C. Wahl, G. Das, and W. Zemke, *Phys. Rev.* **A7**, 69 (1973).
44. W. T. Zemke, G. Das, and A. C. Wahl, *Chem. Phys. Lett.* **14**, 310 (1972).
45. G. Das and A. C. Wahl, *J. Chem. Phys.* **56**, 3532 (1972).
46. J. Berkowitz and A. C. Wahl, *Adv. Flourine Chem.* **7**, 147 (1973).
47. D. Demoulin and M. Jungen, *Chem. Phys.* **16**, 311 (1976).
48. J. S. Cohen and B. Schneider, *J. Chem. Phys.* **61**, 3230 (1974).
49. R. P. Saxon and B. Liu , *J. Chem. Phys.* **64**, 3291 (1976).
50. P. J. Hay, T. H. Dunning, Jr., and R. C. Raffenetti, *J. Chem. Phys.* **65**, 2679 (1976).
51. R. S. Mulliken, *Chem. Phys. Lett.* **25**, 305 (1974).
52. M. Leoni and K. Dressler, *Helv. Phys. Acta* **45**, 959 (1972).
53. H. Lefebvre-Brion and C. M. Moser, *J. Chem. Phys.* **43**, 1394 (1965).
54. W. C. Ermler and R. S. Mulliken, *J. Mol. Spectros.* **61**, 100 (1976).
55. D. C. Cartwright, W. J. Hunt, W. Williams, S. Trajmar, and W. A. Goddard, III, *Phys. Rev.* **A8**, 2436 (1973).
56. R. J. Buenker and S. D. Peyerimhoff, *Chem. Phys.* **8**, 324 (1975).
57. R. S. Mulliken, *Acc. Chem. Res.* **9**, 7 (1976).
58. R. S. Mulliken, *Chem. Phys. Lett.* **14**, 141 (1972), Table 2.
59. R. S. Mulliken, *Chem. Phys. Lett.* **46**, 197 (1977).
60. R. J. Buenker and S. D. Peyerimhoff, *Chem. Phys.* **8**, 325 (1975); *Chem. Phys. Lett.* **34**, 225 (1975); **36**, 415 (1975).
61. K. Dressler, *Can. J. Phys.* **47**, 547 (1969); M. Leoni and K. Dressler, *Helv. Phys. Acta.* **45**, 959 (1972); P. K. Carroll and K. Yoshino, *J. Phys. B* **5**, 1614 (1972).
62. W. C. Ermler, A. D. McLean, and R. S. Mulliken, work in progress.
63. H. H. Michels, unpublished work, see *J. Chem. Phys.* **53**, 841 (1970).
64. R. Gallusser and K. Dressler, to be published.
65. R. J. Buenker and S. D. Peyerimhoff, *Chem. Phys.* **8**, 324 (1975); *Chem. Phys. Lett.* **34**, 225 (1975); **36**, 415 (1975).
66. M. Yoshimine, K. Tanaka, H. Tatawaki, S. Obara, F. Sasaki, and K. Ohno, *J. Chem. Phys.* **64**, 2254 (1976); R. J. Buenker, S. D. Peyerimhoff, and M. Perić, *Chem. Phys. Lett.* **42**, 383 (1976).
67. H. Hellmann, "Einführung in die Quantenchemie," p. 285, Deuticke, Leipzig, 1937; R. P. Feynman, *Phys. Rev.* **56**, 340 (1939).
68. T. Berlin, *J. Chem. Phys.* **19**, 208 (1951).
69. (a) R. F. W. Bader and W. H. Henneker, *J. Am. Chem. Soc.* **87**, 3063 (1965), and later papers.
    (b) R. F. W. Bader and A. K. Chandra, *Can. J. Chem.* **46**, 953 (1968).
70. F. D. Billingsley and M. Krauss, *J. Chem. Phys.* **60**, 2767 (1974).
71. V. P. Gutschick and V. McKoy, *J. Chem. Phys.* **58**, 2397 (1973).
72. M. L. Sink, H. Lefebvre-Brion, and J. A. Hall, *J. Chem. Phys.* **62**, 1802 (1975).
73. E. A. Laws, R. M. Stevens, and W. N. Lipscomb, *J. Chem. Phys.* **54**, 4269 (1971).
74. S. R. Langhoff, *J. Chem. Phys.* **64**, 1245 (1976).

CHAPTER VI

HETEROPOLAR DIATOMIC MOLECULES

A. SCF CALCULATIONS

Diatomic hydrides have been discussed in Chapter IV. Cade and Huo [1a][1] have summarized rather accurate SCF calculations on the ground and some excited states of heteropolar molecules AB, AB^+, and AB^-, where A and B are first-row atoms. As a sample of their reported results, Table 1 shows their basis sets and the ε's and STF compositions of the MOs in CO. Comparison with Table V.2 for the isoelectronic and physically very similar molecule N_2 illustrates the effect of going from a homopolar to a heteropolar molecule. Note the strong polarities of all the MOs, not all in the same direction, but predominantly in favor of O (see also Table 11 and Section F). In spite of this, the dipole moment of CO is nearly zero.

Table 2 is a similar table for the much more strongly heteropolar (nearly pure ionic) molecule LiF, whose structure is considered in detail in Section E.

[1] The authors state that some of the reported results (including those reproduced here in Tables 1 and 2) could be appreciably improved by reoptimization of the ζ's and/or adding further STFs to the basis set. Table 1 is from a paper on CO and BF by Huo [1b]. Considerably more accurate SCF calculations have been made by McLean and Yoshimine [1c] and by Green [1d]: $E = -112.7892$ at R_e.

TABLE 1

STFs for the Ground State of CO ($1\sigma^2 2\sigma^2 3\sigma^2 4\sigma^2 1\pi^4 5\sigma^2$, $^1\Sigma^+$) at Its R_e (2.132 a.u.), Followed by ε Values and STF Coefficients for the Occupied MOs[a–c]

STF	MO: $-\varepsilon$:	1σ 20.661	2σ 11.359	3σ 1.519	4σ 0.802	5σ 0.553
$1s_O$ (7.309)[d]		0.922	−0.001	−0.205	0.102	0.014
$1s_O$ (11.810)		0.083	0.000	−0.004	0.004	0.001
$2s_O$ (1.929)		−0.001	−0.000	0.585	−0.425	−0.068
$3s_O$ (4.333)		0.006	0.001	0.262	−0.140	−0.019
$2p\sigma_O$ (1.488)		−0.000	−0.001	0.098	0.367	−0.208
$2p\sigma_O$ (2.833)		0.000	0.001	0.131	0.358	−0.179
$2p\sigma_O$ (5.898)		0.001	−0.000	0.010	0.040	−0.020
$3d\sigma_O$ (2.448)		0.000	0.000	0.031	0.036	−0.022
$1s_C$ (5.374)		0.000	0.919	−0.121	−0.131	−0.111
$1s_C$ (9.065)		−0.000	0.087	−0.002	−0.006	−0.012
$2s_C$ (1.308)		0.001	−0.000	−0.004	0.111	0.689
$3s_C$ (2.957)		0.000	0.005	0.198	0.265	0.212
$2p\sigma_C$ (1.187)		0.001	−0.001	−0.038	−0.070	−0.216
$2p\sigma_C$ (2.171)		0.000	0.002	0.198	0.132	−0.335
$2p\sigma_C$ (5.821)		0.000	0.001	0.004	0.002	−0.012
$3d\sigma_C$ (2.756)		0.000	0.001	0.029	0.013	−0.021

STF	MO: $-\varepsilon$:	1π 0.638
$2p\pi_O$ (1.473)		0.442
$2p\pi_O$ (2.848)		0.375
$2p\pi_O$ (5.785)		0.045
$3d\pi_O$ (2.383)		0.039
$2p\pi_C$ (1.584)		0.326
$2p\pi_C$ (3.344)		0.053
$3d\pi_C$ (2.067)		0.061
$4f\pi_C$ (2.809)		0.013

[a] Computed energy $E = -112.7860$ a.u.

[b] Note that the z axes of the two atoms are here taken as pointing toward each other. In particular this makes $2p\sigma_A$ and $2p\sigma_B$ have a positive overlap where their coefficients have *like signs*.

[c] This table reproduces Huo's results in Ref. 1b.

[d] ζ Values in parentheses.

TABLE 2

STFs for the Ground State of LiF ($1\sigma^2 2\sigma^2 3\sigma^2 4\sigma^2 1\pi^4$, $^1\Sigma^+$) at R_e (2.955 a.u.), Followed by ε Values and STF Coefficients for the MOs[a,b]

STF		MO: / $-\varepsilon$:	1σ 26.100	2σ 2.428	3σ 1.370	4σ 0.495
$1s_F$	(7.938)[c]		0.955	−0.007	−0.273	0.018
$1s_F$	(14.201)		0.084	0.000	0.006	−0.001
$2s_F$	(2.057)		−0.000	0.015	0.551	−0.058
$2s_F$	(3.332)		0.003	0.011	0.517	−0.035
$3s_F$	(9.962)		−0.042	−0.000	−0.022	0.000
$2p\sigma_F$	(1.612)		−0.000	0.009	0.016	0.576
$2p\sigma_F$	(3.176)		−0.000	0.007	0.016	0.412
$2p\sigma_F$	(6.165)		0.000	0.001	0.001	0.075
$3d\sigma_F$	(1.963)		0.000	0.003	0.004	0.019
$1s_{Li}$	(2.441)		−0.000	0.888	−0.069	−0.112
$1s_{Li}$	(4.601)		−0.000	0.125	−0.006	−0.006
$2s_{Li}$	(1.173)		0.000	0.003	0.036	0.081
$2s_{Li}$	(1.726)		−0.000	−0.005	−0.017	−0.049
$2p\sigma_{Li}$	(1.228)		0.000	0.006	0.034	0.060
$2p\sigma_{Li}$	(1.714)		−0.000	−0.019	0.000	0.028
$3d\sigma_{Li}$	(1.936)		0.000	−0.001	0.009	0.009

STF		MO: / $-\varepsilon$:	1π 0.468
$2p\pi_F$	(1.434)		0.433
$2p\pi_F$	(2.356)		0.370
$2p\pi_F$	(4.249)		0.259
$2p\pi_F$	(9.435)		0.008
$3d\pi_F$	(1.963)		0.017
$2p\pi_{Li}$	(0.831)		0.066
$2p\pi_{Li}$	(1.499)		0.011
$3d\pi_{Li}$	(1.131)		0.030

[a] Computed energy $E = -106.9904$ a.u.
[b] See footnote to Table 1.
[c] ζ Values in parentheses.

SCF plus extensive CM calculations on numerous excited states of CO and on SiO are reported in Section F.

Nonrelativistic SCF computations on PbO using a minimal basis set have been shown [2] to give as good agreement with experiment for spectroscopic constants as similar computations on CO. The computed MOs show strong similarities for the two molecules, suggesting that perhaps for approximate information of chemical interest, the very large relativistic effects in the total energy for heavy molecules is predominantly atomic in nature and can be ignored.

McLean and Yoshimine [1c] have published a table of their SCF calculations on a large number of diatomic and linear polyatomic molecules. The diatomic molecules include both molecules composed of first-row atoms and many containing higher-row atoms. Their table also includes many alkali halide diatomics computed by Matcha [3a]. An SCF calculation on CN^- at five different R values should be mentioned here [4].

For LiCl, NaF, NaCl, and KF, Matcha used an ample basis consisting of free-atom STFs plus $d\sigma$, $f\sigma$, $d\pi$, and $f\pi$ polarization functions; and for KCl, LiBr, RbF, and NaBr a less complete (double-ζ plus polarization) basis set. From the resulting wave functions he has computed dipole moments, quadrupole moments, spectroscopic constants, and the field gradients at each nucleus and from this values of the nuclear quadrupole moments. Matcha [3b] has computed relativistic contributions to binding energy in several alkali halide molecules.

In a very instructive paper, Curtiss *et al.* [5] give contour maps of total density and density difference (compared with separate atoms) functions for a series of alkali halides. (For a related discussion of LiF, see Section E.) They show also the boundaries between binding and antibinding regions, and show how they change with changes in relative nuclear charges at the two centers. For NaF as an example (electron configuration $1\sigma^2 2\sigma^2 3\sigma^2 4\sigma^2 1\pi^4 5\sigma^2 6\sigma^2 2\pi^4$) they give a detailed analysis, including discussions of the individual MOs. Density difference maps as a function of R are also shown. These are interpreted in terms of charge transfer (at R_e, 0.7e from $3s_{Na}$ to $2p\sigma_F$ in 6σ, 0.1e $2p\sigma_{Na}$ to F in 4σ, 0.1e F to Na in 5σ and in 2π).

The transfers in 4σ and 5σ disappear at larger R values, but those in 6σ and 2π remain. One can perhaps conclude that the effect in 6σ is a genuine charge transfer, but that those in 4σ, 5σ, and 2π can be understood as polarization effects. Thus the charge distribution could be described approximately as $Na^{+0.7}F^{-0.7}$ with polarization of the untransferred electrons. An interesting graph showing the spatial distribution of the polarization in the 2π MO is given.

Matcha [3] has calculated the dipole moments for several R values, and from these and spectroscopic information he has computed the dipole moment in each case as a function of the vibrational quantum number v:

$$\mu = u_0 + \mu_1(v+\tfrac{1}{2}) + \mu_2(v+\tfrac{1}{2})^2 + \cdots. \tag{1}$$

Not only μ_0, but also μ_1 is of considerable interest since the intensity of any infrared fundamental ($\Delta v = 1$) is approximately proportional to $\mu_1{}^2$. Likewise μ_2 enters into the intensity for $\Delta v = 2$ bands, and so on. Table 3 compares Matcha's computed μ with experimental values.

The agreements with experiment for these SCF computed μ values are rather good. Better results might be expected if correlated wave functions were available, as can be seen by referring to Table IV.17 on the diatomic hydrides. We note, however, that the SCF-value agreements are much better for these nearly pure ionic molecules than for the less polar hydrides. The same is true for the nearly ionic LiH in the hydride tables. Regarding dipole moments, see also Section F.

SCF and CM calculations on LiO, AlO, FeO, KrF, KrF^+, NaLi, $NaLi^+$, and NO are reported in Section F. SCF calculations on the ground states of ScO, ScF, TiN, TiO, and VO have been made by Carlson *et al.* [6]. A minimal basis set was used for ScO, TiO, and VO, with approximate optimization of orbital exponents for the valence-shell orbitals; plus a few polarization functions. A larger basis set was used for TiN and ScF. In the last-quoted paper, charge-density contours for the valence-shell MOs of TiO and ScF, and for corresponding localized MOs, are given, also contour maps for the total charge densities.

O'Hare and Wahl [7] have made a number of accurate SCF calculations on the ground states of diatomic halides [OF, OF^+, OF^-; SF, SF^+, SF^-, SeF, SeF^+, SeF^-; ClO, ClO^+, ClO^-; NF (including two excited states), NF^+, and NF^-; PF, PF^+, PF^-; CF, CF^+, CF^-, SiF, SiF^+, SiF^-] at or

TABLE 3

Dipole Moments of Alkali Halide Molecules [Eq. (1)]

Molecule	SCF computed			Experimental		
	μ_0	μ_1	μ_2	μ_0	μ_1	μ_2
NaF	8.335	0.0610	0.0002	8.1235	0.0644	0.0004
KF	8.6886	0.0634	−0.0001	8.5583	0.0684	0.0002
$^7Li^{35}Cl$	7.2182	0.0753	0.0002	7.085	0.086	0.0006
NaCl	9.1540	0.0580	0.0003	8.9734	0.0570	0.0005
$^{39}K^{35}Cl$	10.4564	0.0525	0.0001	10.2391	0.0597	−0.0002
$^6Li^{79}Br$	6.9758	0.0638	0.0002	7.2264	0.0832	0.0006

near R_e, and on several properties of these molecules and ions; estimates of dissociation energies, ionization potentials, and electron affinities are reported. SCF calculations on LiB, LiAl, BeHe, and BeAr have been made by Kaufman *et al.* [8].

SCF calculations on alkali metal oxides are also available [9].[2]

For the isoelectronic molecules C_2, BN, and BeO, for a long time there were doubts as to whether the ground state is $\ldots 3\sigma^2 4\sigma^2 1\pi^3 5\sigma$, $^3\Pi$ or $\ldots 3\sigma^2 4\sigma^2 1\pi^4$, $^1\Sigma^+$. For C_2, it was finally established that, by a small margin, $^1\Sigma_g{}^+$ is the ground state. For BN, an SCF calculation by Verhaegen *et al.* makes $^3\Pi$ the ground state by a wide margin, but the greater estimated correlation energy of $^1\Sigma^+$ brings it down to about 0.6 eV above $^3\Pi$ [10]. This and the fact that a triplet spectrum of BN is observed spectroscopically make it probable that the ground state is $^3\Pi$, but it would be desirable that this conclusion should be checked by a careful CM calculation. In the case of BeO, although SCF calculations make $^3\Pi$ the ground state, the situation has been definitely resolved in favor of $^1\Sigma^+$ by CM calculations of Schaefer *et al.* discussed in Section F. BeO is of especial interest as the first member of the alkaline earth oxide family.

Experimentally, a $^1\Pi$ state which must be $\ldots 3\sigma^2 4\sigma^2 1\pi^3 5\sigma$ is 9506 cm^{-1} above the known $^1\Sigma^+$ state in BeO. SCF calculations by Huo *et al.* locate the corresponding $^3\Pi$ state only 923 cm^{-1} lower [11]. The combined evidence corroborates the results of the CM computations of Schaefer *et al.* In both the $^1\Sigma^+$ and the Π states, the structure according to the SCF computations is approximately Be^+O^-, with the 4σ a Be^+O^- bonding MO and 5σ and 1π both approximately Be atom $2p\sigma$ and $2p\pi$ AOs (see Ref. 11 for contour maps of these). Next above the $^1\Pi$ is probably $\cdots 4\sigma 5\sigma 1\pi^4$, $^3\Sigma^+$. A point of interest is that the $^1\Sigma^+$ cannot dissociate to unexcited atoms (Be ^1S plus O ^3P), but one atom must be excited on dissociation.

For MgO the situation is similar to that in BeO, except that the A $^1\Pi$ state is experimentally only 3563 cm^{-1} above the $^1\Sigma^+$. However, SCF calculations by Schamps and Lefebvre-Brion with a double-ζ basis set show that the experimentally unknown $^3\Pi$ state is only about 1130 cm^{-1} below A $^1\Pi$, and thus in MgO, X $^1\Sigma^+$ is again indicated as the ground state [12]. Approximate CM calculations show considerable mixing in this state of $\ldots 6\sigma 7\sigma 2\pi^4$, $^1\Sigma^+$ into $\ldots 6\sigma^2 2\pi^4$. The same authors have made SCF calculations on some of the higher excited states of MgO, and on the $\ldots 6\sigma 2\pi^4$, $^2\Sigma^+$ and $\ldots 6\sigma^2 2\pi^3$ states of MgO^+, and find the $^2\Pi$ state to be the ground state, at least by SCF.

A similar discussion and SCF calculation using an ample basis set including polarization functions has been carried out by Carlson *et al.* for

[2] On LiO, see Section F.

CaO [13]. Because of the low excitation energy of the d electrons in the Ca atom, somewhat different behavior in CaO than in BeO and MgO would not be surprising; corresponding remarks apply to SrO and BaO. Just as for BeO and MgO, the SCF calculations make $^3\Pi$ ($\ldots 8\sigma^2 9\sigma 3\pi^3$) the ground state, but other evidence favors $^1\Sigma^+$ (see Carlson *et al.* [13] for a review). Measurements of molecular-beam resonance spectra appear to be reasonably conclusive in establishing the spectroscopic X $^1\Sigma^+$ state as the ground state in the cases of SrO and BaO. The work of Schaefer *et al.* on BeO (see Section F) while not conclusive as to the heavier oxides, strongly favors $^1\Sigma^+$ as the ground state in all. In the case of CaO, one probable conclusion from the calculations is that the lowest $^1\Sigma^+$ state is predominantly $\ldots 8\sigma 9\sigma 3\pi^4$ but mixed with a considerable amount of $\ldots 8\sigma^2 3\pi^4$. Earlier calculations of Yoshimine on the alkaline earth oxides deal only with the lowest $^1\Sigma^+$ state [14, 1c]. He finds that the dipole moment goes through a maximum at an R value near R_e in all, indicating a Ca^+O^- structure near R_e. Dissociation, however, must be to neutral atoms.

In Section V.B the relations between orbital energies and ionization potentials were discussed with special reference to O_2 and N_2. A similar discussion of CO has been given [15]. Ionization of a 1σ CO electron gives a $1s_O$ hole state of CO^+, that of 2σ gives a $1s_C$ hole state. Using fairly good SCF computations, the validity of Koopmans' theorem and related matters are discussed, using three-dimensional perspective plots of electron density and density difference functions. Comparisons are made between the 1s hole state and CF^+, and between the $1s_C$ hole state and NO^+.

SCF calculations were made some time ago on various hyperfine structure coupling constants of NO, with results agreeing reasonably well with experiments [16]. More accurate calculations would now be possible.

SCF calculations on the quadrupole moment of NO^+ show reasonable agreement with experiment [17].

Boyd [18] has presented instructive total and difference density maps obtained from SCF computations on MgO, PO, and PN.

B. MOLECULAR MULTIPLETS

Important information about molecular structure can be obtained from the widths of molecular multiplets [19]. Usually the interaction between the spin of each electron and its own magnetic field predominates over other smaller terms, and in that approximation one has for the multiplet energy Γ of a diatomic molecule,

$$\Gamma = \sum \gamma_i = \int \Psi^* \left(\sum_i a_i \, l_i \cdot s_i \right) \Psi \, dv = A\Lambda\Sigma. \tag{2}$$

For the case of a single electron outside closed shells, $A\Lambda\Sigma$ reduces to $a\lambda\sigma$, where λ is the orbital and σ the spin quantum number of the electron in its MO (Λ and Σ in general are the corresponding *total* quantum numbers). Two important special cases are $\ldots\pi$, ${}^2\Pi$ and $\ldots\pi^3$, ${}^2\Pi$, each with two components ${}^2\Pi_{1/2}$ ($\Lambda = 1$, $\Sigma = -\frac{1}{2}$) and ${}^2\Pi_{3/2}$ ($\Lambda = 1$, $\Sigma = +\frac{1}{2}$). In the first of these, $A = a$ (regular doublet, ${}^2\Pi_{1/2}$ below ${}^2\Pi_{3/2}$) while in the second $A = -a$ (inverted doublet, ${}^2\Pi_{1/2}$ above ${}^2\Pi_{3/2}$). For each π MO, $\lambda = 1$, $\sigma = \pm\frac{1}{2}$.

Empirical values of a and A for atomic states can be derived from atomic spectra, and show interesting relations to those of the multiplets of molecules built from these atoms. For a molecule AB, one can easily obtain a somewhat crude relation between atomic and molecular a values, as follows: Let a_{op} be the quantum-mechanical operator which yields a for an atom, then for an electron in an MO of the form $\alpha\chi_{\text{A}}+\beta\chi_{\text{B}}$, where χ_{A} and χ_{B} are MAOs [cf. Eq. (II.3)],

$$a_{\text{AB}}\,\lambda\sigma = \int(\alpha\chi_{\text{A}}+\beta\chi_{\text{B}})^*(a_{\text{op}}\,l\cdot s)(\alpha\chi_{\text{A}}+\beta\chi_{\text{B}})\,dv. \tag{3}$$

Now the MAOs χ_{A} and χ_{B} are not orthogonal and so ${\chi_{\text{A}}}^2+{\chi_{\text{B}}}^2 \neq 1$, but $\alpha^2+\beta^2+2\alpha\beta S = 1$. However, let $\alpha' = \alpha/(1-2\alpha\beta S)$ and $\beta' = \beta/(1-2\alpha\beta S)$; then $\alpha'^2+\beta'^2 = 1$. The following, perhaps rather rough approximation, then becomes reasonable:

$$a_{\text{AB}} = (\alpha'^2 a_{\text{A}}+\beta'^2 a_{\text{B}}). \tag{4}$$

This relation could be tested using computed *ab initio* expressions for the MAOs χ_{A} and χ_{B}, but this has apparently not yet been done. However, inspection of empirical spectroscopic atomic and diatomic a's indicates [19] that a relation of this kind is at least qualitatively true.

Extensive spectroscopic data exist on the diatomic alkaline earth halides. The ground state is definitely ${}^2\Sigma^+$ ($\ldots3\sigma^2 4\sigma^2 1\pi^4 5\sigma$) in BeF. BeF is isoelectronic with BO, CO^+, CN, and ${N_2}^+$, and like these, its first excited state is a ${}^2\Pi$. For all but BeF, this has long been known to be an inverted ${}^2\Pi$, showing that the electron configuration is $\ldots3\sigma^2 4\sigma^2 1\pi^3 5\sigma^2$. For BeF, however, the possibility of $\ldots3\sigma^2 4\sigma^2 1\pi^4 2\pi$ with ${}^2\Pi$ regular must also be considered. Here we note that a_{Be} is 1 cm^{-1}, while a_{F} is 271 cm^{-1}. Since SCF calculations [20] show that BeF is highly polar (essentially Be^+F^-), the 1π MO has $\beta' \gg \alpha'$ in Eq. (4), and a in the $\ldots1\pi^3 5\sigma^2$ state (here $A = -a$) should be now a_{F}. But in the $\ldots1\pi^4 2\pi$ state, the 2π MO has $\beta' \ll -\alpha'$, and $A = a$ should be small (but not nearly as small as a_{Be}, since even a small β' is multiplied here by a large a_{F}). However, Λ-doubling evidence in the fine structure of the band lines seemed to indicate that the observed ${}^2\Pi$ is inverted, although with a numerically small A (-16.46) [19]. This result could be accounted for if the observed state is predominantly $\ldots1\pi^4 2\pi$

with a small positive A but with some configuration mixing with $\ldots 1\pi^3 5\sigma^2$ with a large negative A.

The matter has apparently been settled in a two-pronged attack by Walker and Richards:

(a) they have made good SCF calculations [20] on the ground and the two $^2\Pi$ states, and find good agreement for the $\ldots 1\pi^4 2\pi$ with the spectroscopic constants of the observed state; also, in another paper [21], they conclude that CM between the two $^2\Pi$ states is essentially negligible, and so could not account for an inverted observed state;

(b) they have shown by SCF calculations that the observed Λ-type doubling in BeF and MgF can be accounted for by a regular instead of an inverted $^2\Pi$ state.

The final conclusion is that the *observed* $^2\Pi$ in BeF is *regular*, with $A = +21.8$ cm^{-1} (computed [19], 23.7 cm^{-1}). In the analogous case of MgF, where the Λ-doubling has also been troublesome, they conclude that the observed $^2\Pi$ is a regular with $A = +37.3$ cm^{-1} [21, 22].

Extensive experimental data exist on $^2\Pi$ states of the heavier alkaline earth halides (including chlorides, bromides, and iodides). These can be interpreted in terms of Eq. (4), but probably only if we consider the possibility that the χ_A for the alkaline earth metal in Eq. (3) is not just $np\pi$, but partly or largely $nd\pi$. For example [19a], A for $^2\Pi$ of CaCl (76 cm^{-1}) is smaller than a for np either of Ca (106 cm^{-1}) or of Cl (587 cm^{-1}), whereas if a were mostly 4pπ of Ca but with a little 3pπ of Cl, as expected for a $\pi^4\pi$ configuration, A should be somewhat *larger* than a for the Ca atom. Participation for Ca of 3dπ, with its smaller a, instead of 4pπ, can account for the low value of A. However, more experimental work seems needed.

Much more accurate than Eq. (4), Walker and Richards have now made SCF calculations on the spin-orbit coupling coefficients of the low-energy $^2\Pi$ states of a number of molecules (BeH, BH^+, CN, NH^-, OH, HF^+, MgH, AlH^+, SiH, PH^-, SH, BO (two states), CO^+ (two states), NO, $N_2{}^+$, PO, ClO), and the low $^3\Pi$ states of CO and C_2, and they obtain very good agreements between observed and calculated values [23a]. They use the Dirac Hamiltonian for the spin-orbit coupling. They have also dealt with Λ doubling [23b].

C. CHARGE DISTRIBUTIONS

Contour maps of charge densities, and of difference densities as compared with those of separated atoms, have been discussed for diatomic hydrides in Section IV.B (q.v.), also for homopolar diatomics in Section V.D. Figures 1 and 2 show instructive comparisons of overall and difference densities

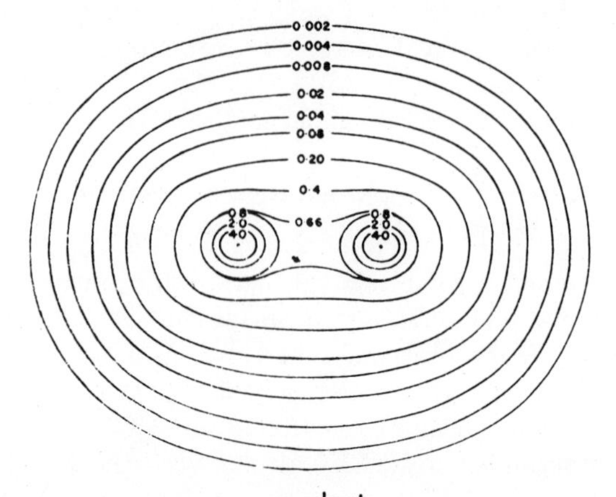

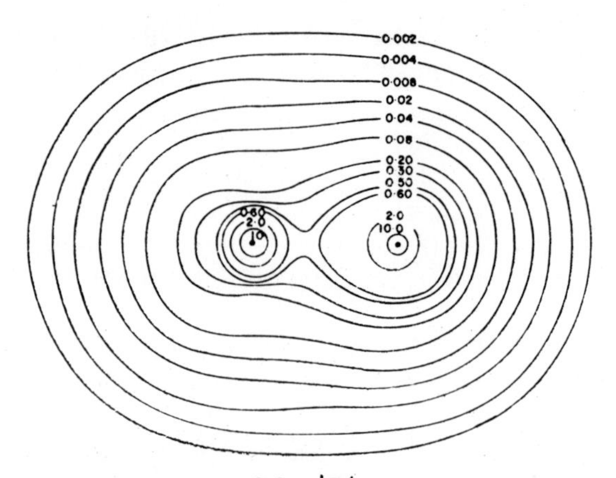

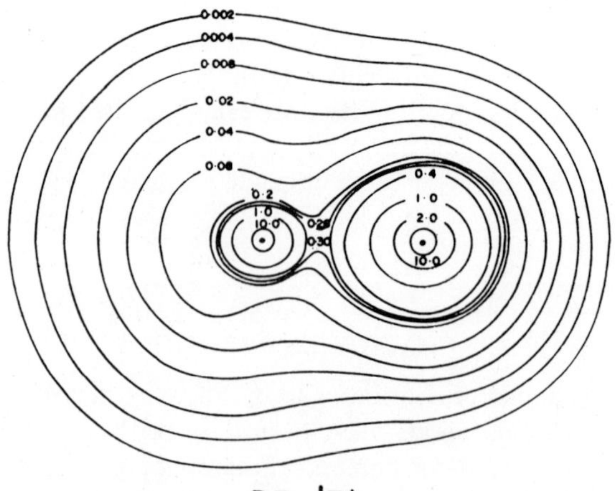

FIG. 1 Contour maps of the total molecular charge distributions. [From R. F. W. Bader and A. D. Bandrauk, *J. Chem. Phys.* **49**, 1953 (1968).]

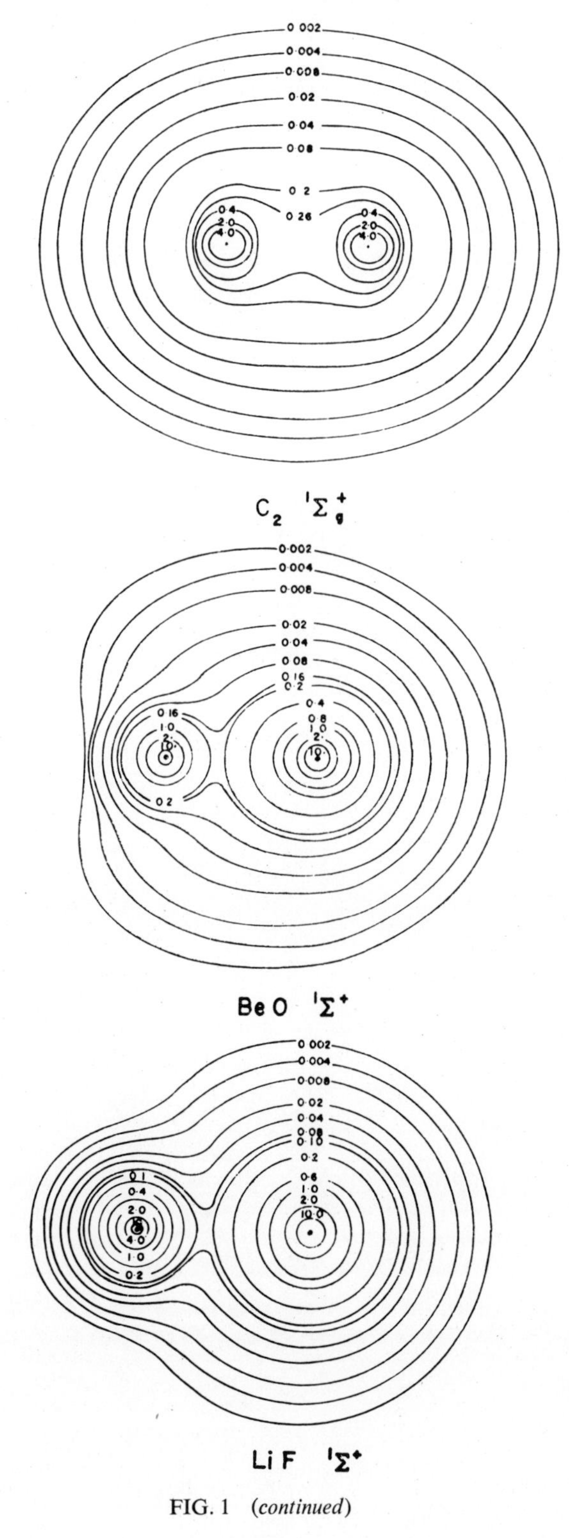

FIG. 1 (*continued*)

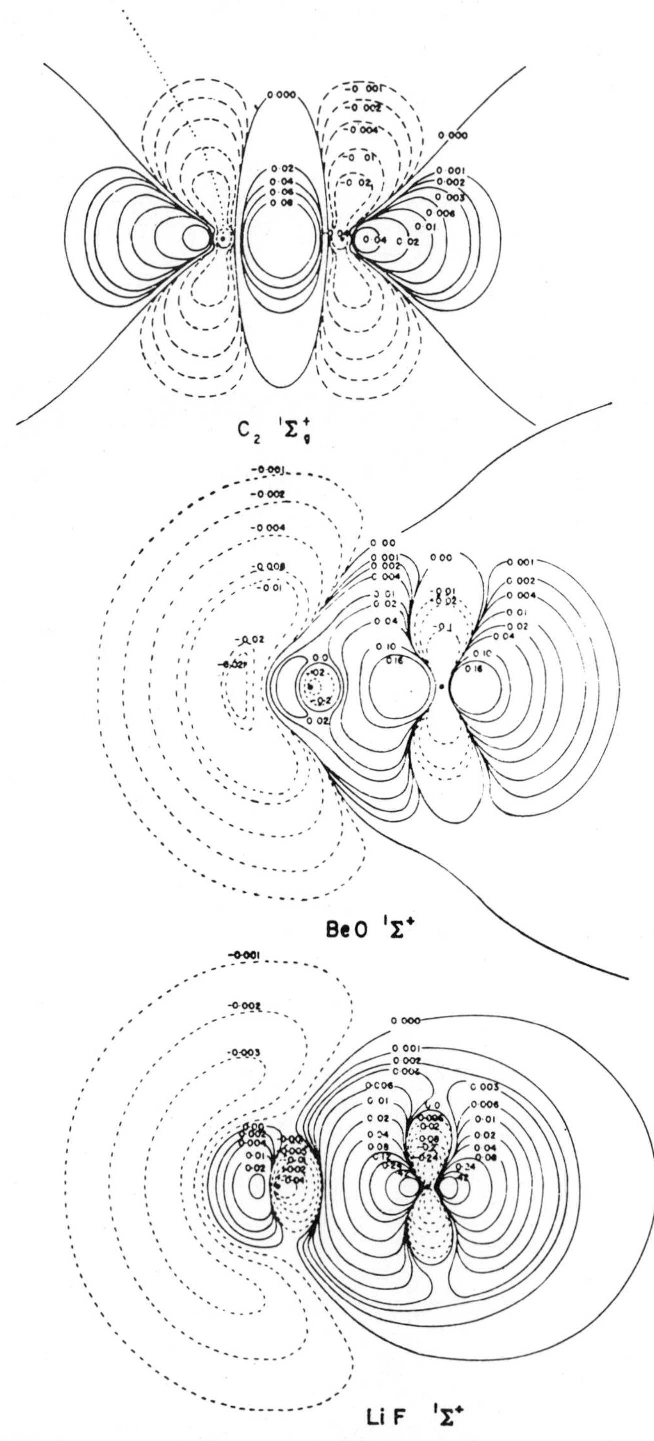

FIG. 2 Contour maps of the density difference distributions. [From. R F. W. Bader and A. D. Bandrauk, *J. Chem. Phys.* **49**, 1953 (1968).]

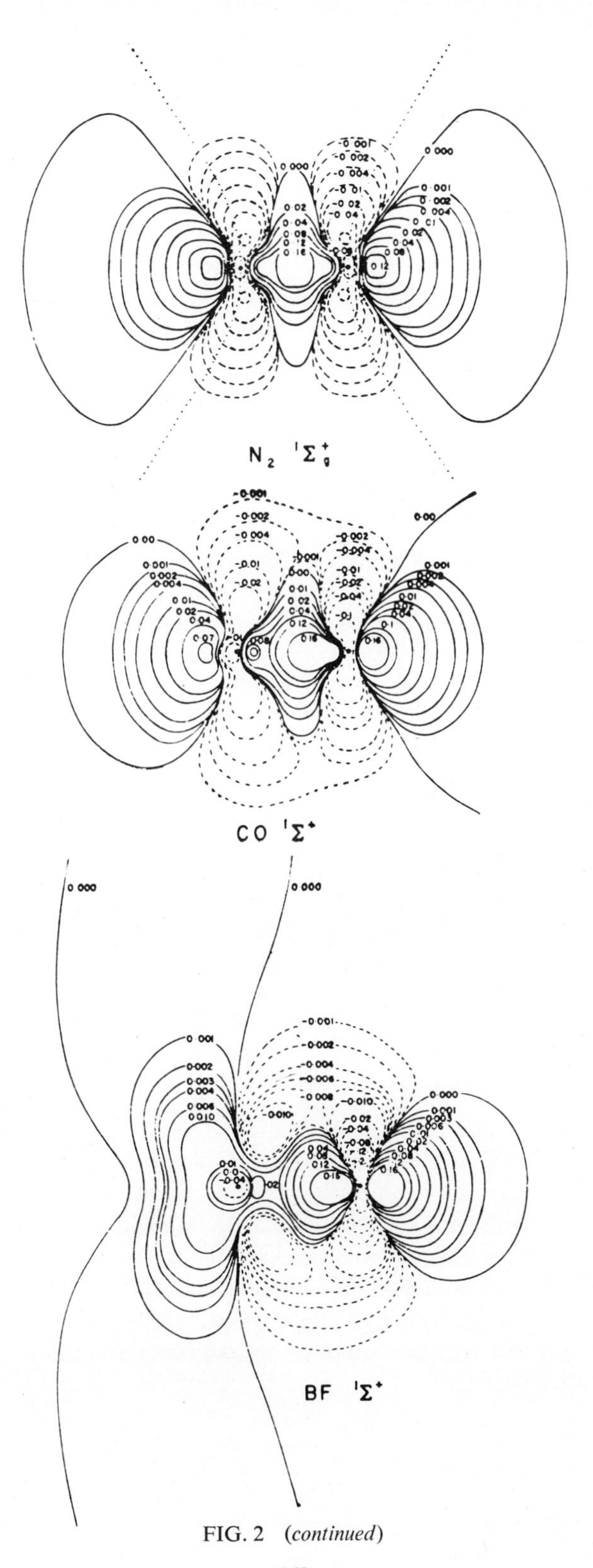

FIG. 2 (*continued*)

respectively, for each of the two isoelectronic series N_2, CO, BF and C_2, BeO, and LiF.[3] The changes with increasing polarity are striking.

One thing of interest which can be read from the overall density maps is a measure of the dimensions of a molecule. By using the 0.002 density contour as a gauge, lengths and widths of the molecules just mentioned are obtained as given in Table 4. This contour roughly defines the van der Waals dimensions of the molecule, and contains within it over 95% of the total charge density [25].[4] If L is the length as just defined of a molecule AB, and r_A and r_B are van der Waals radii of its atoms *in the molecule*, then $L = r_A + R_e + r_B$. Table 4 includes comparisons with the r_A and r_B of the free atoms. A similar table for first-row diatomic hydrides has also been published [26].

TABLE 4

Dimensions of Some First-Row Molecules

Molecule AB	R_e (a.u.)	Length (a.u.)	Width (a.u.)	r_A Mol. (Atom)	r_B Mol. (Atom)
N_2	2.07	8.2	6.4	3.1 (3.0)	3.1 (3.0)
CO	2.13	8.5	6.2	3.4 (3.2)	2.8 (2.9)
BF	2.39	9.0	6.5	3.8 (3.4)	2.8 (2.8)
C_2	2.35	8.5	7.0	3.1 (3.2)	3.1 (3.2)
BeO	2.52	6.9	6.7	1.3 (3.6)[a]	3.1 (2.9)
LiF	2.96	7.6	6.0	1.7 (3.3)[a]	3.0 (2.8)

[a] In these cases *ionic radii* (for Be^+, Li^+) would be more appropriate for the molecules. For the free ions, these are 1.4 and 1.8 a.u., respectively.

D. POPULATION ANALYSIS AND BONDING

Tables 5 and 6 give some details of the overlap populations in the MOs of CO and LiF. Tables 7–9 compare the MO and total overlap populations in the molecules of each of three first-row isoelectronic series; the overlaps are seen to decrease with increasing polarity in each series. Table 10 tests the correlation of n and $n\bar{I}$ with dissociation energies; it is seen that the latter do not fall off as rapidly as n and $n\bar{I}$. This fact is understandable because as the molecule becomes more polar, although covalent bonding (measured by n or $n\bar{I}$) falls off, ionic bonding (perhaps approximately proportional to Q/R_e) builds up.

[3] Bader and Bandrauk also show instructive density profiles along the internuclear axis.
[4] Reference 25 also includes width and length data on first-row homopolar diatomics.

TABLE 5

Overlap Populations for Ground State $1\sigma^2 2\sigma^2 3\sigma^2 4\sigma^2 1\pi^4 5\sigma^2$, $^1\Sigma^+$ of CO at R_e

MO	$n_i(\text{s, s})$	$n_i(\text{s}_\text{C}, \text{p}\sigma_\text{O})$	$n_i(\text{s}_\text{O}, \text{p}\sigma_\text{C})$	$n_i(\text{p}\sigma, \text{p}\sigma)$	$n_i(\text{d}\sigma, \text{all})$	$n_i(\sigma, \sigma)$	$n_i(\text{p}\pi, \text{p}\pi)$	$n_i(\text{d, f}\pi, \text{all})$	$n_i(\pi, \pi)$	
1σ	0.000	−0.000	0.000	0.000	0.000	0.000				
2σ	−0.000	−0.000	0.000	0.000	0.000	−0.000				
3σ	0.188	0.059	0.160	0.047	0.039	0.492				
4σ	−0.300	0.383	−0.017	0.081	0.014	0.160				
5σ	−0.135	−0.479	0.082	0.196	−0.001	−0.338				
Total σ:	−0.247	−0.037	0.224	0.323	0.052	0.315				
1π							0.697	0.229	0.927	
Total n:										1.242

TABLE 6

Overlap Populations for Ground State $1\sigma^2 2\sigma^2 3\sigma^2 4\sigma^2 1\pi^4$ of LiF at R_e

MO	$n_i(\text{s, s})$	$n_i(\text{s}_\text{Li}, \text{p}\sigma_\text{F})$	$n_i(\text{s}_\text{F}, \text{p}\sigma_\text{Li})$	$n_i(\text{p}\sigma, \text{p}\sigma)$	$n_i(\text{d}\sigma, \text{all})$	$n_i(\sigma, \sigma)$	$n_i(\text{p}\pi, \text{p}\pi)$	$n_i(\text{d}\pi, \text{all})$	$n_i(\pi, \pi)$	
1σ	0.000	−0.000	0.000	−0.000	0.000	0.000				
2σ	0.005	0.007	−0.000	−0.000	0.003	0.013				
3σ	0.012	−0.000	0.051	0.001	0.011	0.074				
4σ	−0.003	−0.023	−0.011	0.086	0.005	0.054				
Total σ:	0.014	−0.017	0.040	0.087	0.018	0.142				
1π							0.131	0.086	0.217	
Total n:										0.359

TABLE 7

Overlap Populations in the $1\sigma^2 2\sigma^2 3\sigma^2 4\sigma^2 1\pi^4$, $^1\Sigma^+$ Isoelectronic Series C_2, BN, BeO, and LiF, at Their R_e Values

	$n(1\sigma)$	$n(2\sigma)$	$n(3\sigma)$	$n(4\sigma)$	$n(\sigma)$	$n(1\pi)$	n
C_2	0.002	−0.002	0.823	−0.452	0.371	1.186	1.557
BN[a]							
BeO	0.000	0.019	0.236	0.158	0.413	0.767	1.180
LiF	0.000	0.013	0.074	0.054	0.142	0.217	0.359

[a] Unfortunately no calculations are available for this state of BN.

TABLE 8

Overlap Populations in the $1\sigma^2 2\sigma^2 3\sigma^2 4\sigma^2 1\pi^4 5\sigma$, $^2\Sigma^+$ Isoelectronic Series, at Their R_e Values

	$n(1\sigma)$	$n(2\sigma)$	$n(3\sigma)$	$n(4\sigma)$	$n(5\sigma)$	$n(1\pi)$	n
N_2^+	0.001	0.000	0.803	0.222	−0.048	1.056	2.035
CN	0.000	0.000	0.664	0.047	−0.080	1.101	1.733
CO^+	0.000	0.001	0.515	0.046	−0.075	1.007	1.494
BO	0.000	0.006	0.381	0·195	−0·093	0.790	1.279
BF^+	0.000	0.003	0.197	0.207	−0.100	0.329	0.636
BeF	0.000	0.006	0.108	0.127	−0.052	0.263	0.452

Tables 11 and 12 give details of the gross atomic populations in CO and LiF respectively. Table 13 is of particular interest in showing how s promotion (s, p hybridization) in both A and B and σ charge transfer from A to B change with increasing polarity in each of the isoelectronic series already mentioned. Table 13 and especially Table 14 show how atomic populations are changed when molecules are formed, including charge transfer with resulting atomic charges Q. The breakdown in Table 14 in terms of σ and π populations shows how these change in molecule formation; the corresponding partial σ and π charges $Q(\sigma)$ and $Q(\pi)$, mostly of opposite sign, are especially interesting.

Charge-density contour maps for the individual valence-shell MOs of CO are shown in Figs. 3–6. These are from a paper by Huo [1b] who also shows similar maps for BF. The maps for CO, as well as the coefficients in Table 1, show that the MOs are strongly polarized or localized. This fact is brought out even more clearly by the gross atomic populations in Table 11, which indicate that the 3σ and 4σ MOs are very largely localized on the oxygen, while the 5σ is almost a pure carbon lone pair MO. On the other hand, about three-fourths of the 1π population is on the O atom.

TABLE 9

Overlap Populations in the $1\sigma^2 2\sigma^2 3\sigma^2 4\sigma^2 1\pi^4 5\sigma^2$, $^1\Sigma^+$ Isoelectronic Series N_2, CO, and BF, at Their R_e Values

	$n(1\sigma)$	$n(2\sigma)$	$n(3\sigma)$	$n(4\sigma)$	$n(5\sigma)$	$n(1\pi)$	n
N_2	0.002	0.006	0.847	0.148	−0.237	1.156	1.922
CO	0.000	−0.000	0.492	0.160	−0.338	0.927	1.242
BF	0.000	−0.000	0.185	0.207	−0.248	0.286	0.430

TABLE 10

Overlap and Bonding in First-Row Heteropolar Molecules

Molecule	n	$\bar{I}$ (eV)	$n\bar{I}$ (eV)	D_e (eV)	R_e (a.u.)
C_2	1.557	11.26	17.53	6.2	2.3481
BN					
BeO	1.180	11.47	13.53	4.6	2.5149
LiF	0.359	11.41	4.10	5.91	2.955
N_2^+	2.035	22.08		8.72	2.113
CN	1.733	12.90	22.36	7.86	2.214
CO^+	1.494	23.21		8.35	2.107
BO	1.279	10.96	14.02	8.28	2.275
BF^+	0.636	21.64		5.0	2.391
BeF	0.452	13.37	6.04	6.0	2.572
N_2	1.922	14.54	27.95	9.76	2.068
CO	1.242	12.44	15.45	11.09	2.132
BF	0.430	12.86	5.53	7.8	2.391
O_2^+	1.157	24.38		6.67	2.122
NO	0.726	14.13	10.26	6.50	2.1747
NF^+	0.411	24.76		—	2.489
CF	0.355	14.34	5.09	5.4	2.402

The C atom localization of the 5σ, and its relatively low ionization energy, explain why CO is an effective electron donor in molecular complexes such as $OC \cdot BH_3$. This characteristic together with its behavior as a π acceptor (acceptation into the unoccupied 2π MO) explain why it is so effective in forming two-way (donor *and* acceptor) charge-transfer complexes such as the carbonyls. Here the high C^+O^- polarity of the 1π MO has its complement, by virtue of the required orthogonality of the two, in an equally strong C^-O^+ polarity in the 2π MO. This polarity strongly favors acceptance into 2π simultaneously with donation from 5σ.

Table 12 shows that the 3σ and 4σ MOs in LiF belong almost completely

TABLE 11

Gross Atomic Populations in Ground State of CO at R_e

MO	$N_i(s_C)$	$N_i(s_O)$	$N_i(p\sigma_C)$	$N_i(p\sigma_O)$	$N_i(d\sigma_C)$	$N_i(d\sigma_O)$	$N_i(\sigma_C)$	$N_i(\sigma_O)$
1σ	0.000	2.000	0.000	0.000	0.000	0.000	0.000	2.000
2σ	2.000	−0.000	0.000	−0.000	0.000	0.000	2.000	−0.000
3σ	0.206	1.453	0.164	0.154	0.016	0.007	0.386	1.614
4σ	0.299	0.411	0.049	1.231	−0.000	0.010	0.348	1.652
5σ	1.165	−0.012	0.702	0.143	0.004	−0.003	1.872	0.128
Total σ:	3.671	3.851	0.916	1.528	0.020	0.015	4.607	5.393
	$N_i(p\pi_C)$	$N_i(p\pi_O)$	$N_i(d\pi_C)$	$N_i(d\pi_O)$	$N_i(\pi_C)$	$N_i(\pi_O)$	$N(C)$	$N(O)$
1π	0.904	2.956	0.109	0.030	1.014	2.987		
Total:							5.621	8.380

TABLE 12

Gross Atomic Populations in Ground State of LiF at R_e

MO	$N_i(s_{Li})$	$N_i(s_F)$	$N_i(p\sigma_{Li})$	$N_i(p\sigma_F)$	$N_i(d\sigma_{Li})$	$N_i(d\sigma_F)$	$N_i(\sigma_{Li})$	$N_i(\sigma_F)$
1σ	0.000	2.000	0.000	0.000	0.000	0.000	0.000	2.000
2σ	1.992	0.003	0.000	0.004	−0.000	0.001	1.992	0.008
3σ	0.016	1.947	0.028	0.002	0.006	−0.000	0.050	1.950
4σ	0.016	0.008	0.053	1.920	0.003	0.000	0.072	1.928
Total σ:	2.024	3.959	0.081	1.926	0.009	0.002	2.114	5.886
	$N_i(p\pi_{Li})$	$N_i(p\pi_F)$	$N_i(d\pi_{Li})$	$N_i(d\pi_F)$	$N_i(\pi_{Li})$	$N_i(\pi_F)$	$N(Li)$	$N(F)$
1π	0.090	3.862	0.045	0.003	0.135	3.865		
Total:							2.249	9.751

TABLE 13

Gross Atomic Populations by l Species in Molecules AB[a]

Molecule	$N(s_A)$	$N(p_A)$	$N(d_A, f_A)$	$N(s_B)$	$N(p_B)$	$N(d, f_B)$	N
C_2	3.526 (4)	2.399 (2)	0.075	3.526 (4)	2.399 (2)	0.075	12
BN	(4)	(1)		(4)	(3)		12
BeO	2.151 (4)	0.825 (0)	0.064	3.899 (4)	5.035 (4)	0.027	12
LiF	2.024 (3)	0.171 (0)	0.054	3.959 (4)	5.788 (5)	0.005	12
CN	3.102 (4)	2.375 (2)	0.073	3.818 (4)	3.572 (3)	0.060	13
BO	2.875 (4)	1.325 (1)	0.066	3.858 (4)	4.835 (4)	0.042	13
BeF	2.794 (4)	0.512 (0)	0.055	3.952 (4)	5.674 (5)	0.014	13
N_2	3.642 (4)	3.258 (3)	0.099	3.642 (4)	3.258 (3)	0.099	14
CO	3.671 (4)	1.820 (2)	0.130	3.851 (4)	4.484 (4)	0.045	14
BF	3.698 (4)	0.745 (1)	0.056	3.929 (4)	5.541 (5)	0.032	14
NO	3.741 (4)	2.891 (3)	0.085	3.857 (4)	4.358 (4)	0.068	15
CF	3.773 (4)	1.772 (2)	0.081	3.938 (4)	5.396 (5)	0.039	15

[a] In parentheses, $N(l)$ for free atom.

TABLE 14

Gross Atomic Populations and Charges by λ Species in Molecules AB[a]

Molecule	$N(\sigma_A)$	$N(\pi_A)$	$N(\sigma_B)$	$N(\pi_B)$	$Q(\sigma_A)$	$Q(\pi_A)$	$Q(A)$	$Q_P(A)$[b]
C_2	4.000 (4)	2.000 (2)	4.000 (4)	2.000 (2)	0.000	0.000	0.000	0.00
BN	(4)	(1)	(4)	(3)				
BeO	2.328 (4)	0.712 (0)	5.672 (4)	3.289 (4)	1.67	−0.712	0.960	0.59
LiF	2.114 (3)	0.135 (0)	5.886 (5)	3.865 (4)	0.886	−0.135	0.751	0.52
CN	3.817 (5)	1.733 (1)	5.183 (4)	2.267 (3)	1.183	−0.733	0.450	
BO	3.475 (5)	0.790 (0)	5.525 (4)	3.210 (4)	1.525	−0.790	0.735	
BeF	3.190 (4)	0.171 (0)	5.810 (4)	3.829 (5)	0.810	−0.171	0.639	
N_2	5.000 (5)	2.000 (2)	5.000 (5)	2.000 (2)	0.000	0.000	0.000	0.00
CO	4.607 (5)	1.014 (1)	5.393 (5)	2.987 (3)	0.393	−0.014	0.379	0.14
BF	4.294 (5)	0.204 (0)	5.706 (5)	3.796 (4)	0.706	−0.204	0.501	0.10
NO	4.766 (5)	1.950 (2)	5.234 (6)	3.050 (2)	0.234	0.050	0.284	0.11
CF	4.408 (5)	1.219 (1)	5.592 (6)	3.781 (3)	0.592	−0.219	0.373	

[a] In parentheses, $N(\lambda)$ for free atom.
[b] Q according to Politzer (private communication).

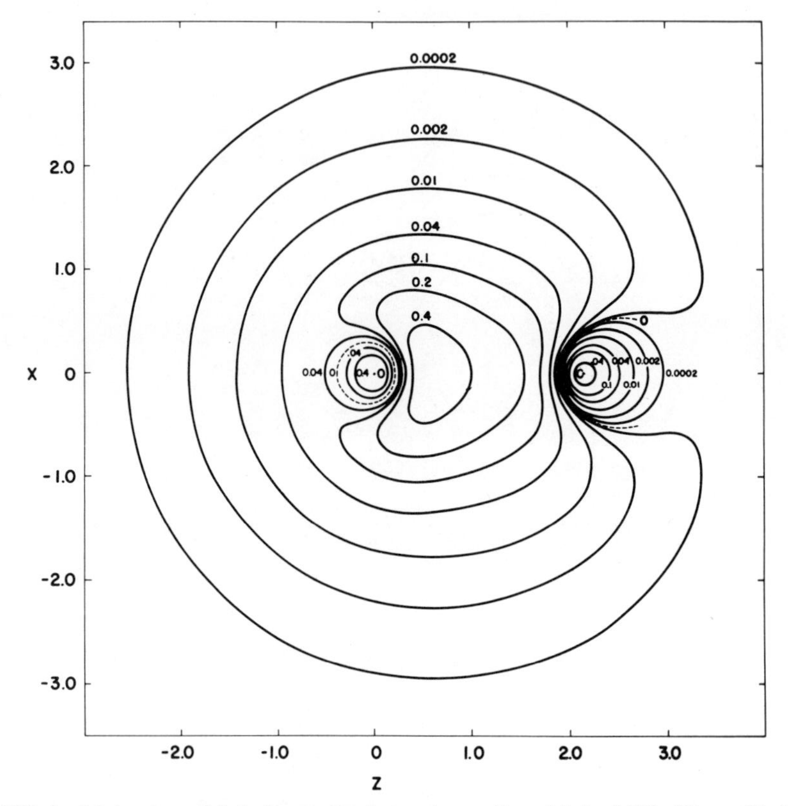

FIG. 3 Molecular orbital charge density contours; 3σ orbital of CO. [From W. M. Huo, *J. Chem. Phys.* **43**, 624 (1964).]

to the F atom, while the 1π is somewhat more shared with the Li atom. The overall population totals give a charge distribution $Li^{+0.75}F^{-0.75}$ in accord with the usual view of LiF as an ion-pair molecule. The Politzer charges (for further discussion, see Section E) in Table 14 for BeO, LiF, and perhaps BF seem questionable in view of our Q values and other evidence for near-ionic character in these molecules.

The overlap populations for CO in Table 5 indicate that most of the σ bonding is concentrated in the 3σ MO, in spite of the fact that the $3\sigma^2$ shell is predominantly a $2s^2$ closed shell (see Table 11). This result is in line with the general tendency of σ bonding in SCF approximation to concentrate in the lowest valence-shell MO, as already discussed for diatomic hydrides in Chapter IV.C, noting, however, that this conclusion may be strongly modified when one contemplates MCSCF MOs.

In its very strong π bonding, as in its physical properties generally, CO

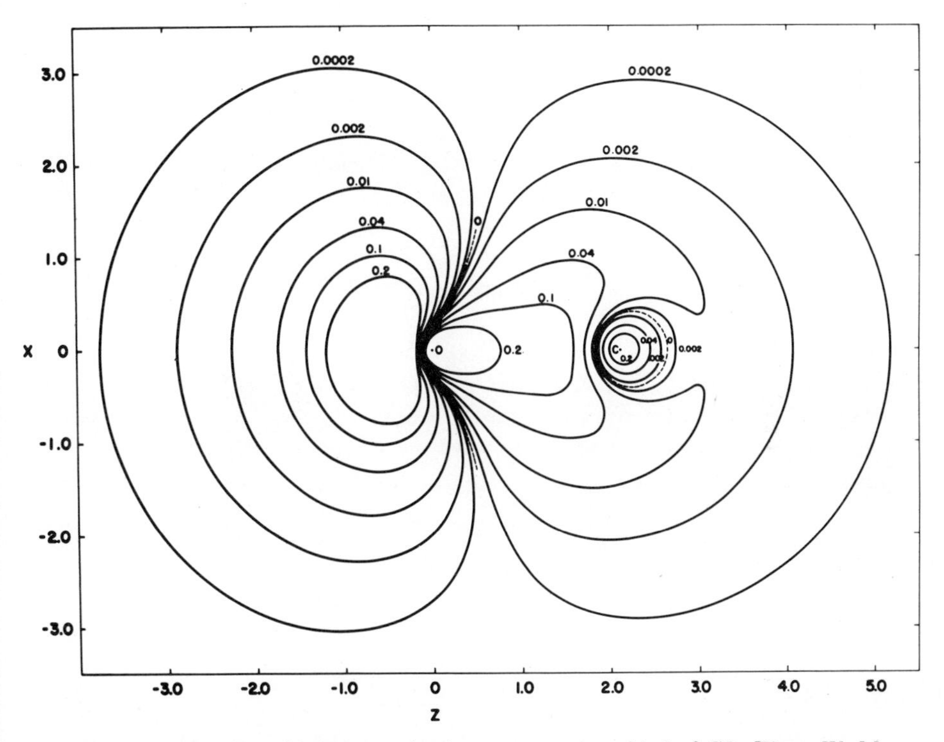

FIG. 4 Molecular orbital charge density contours; 4σ orbital of CO. [From W. M. Huo, *J. Chem. Phys.* **43**, 624 (1964).]

closely resembles N_2 (cf. Section V.E) in spite of the strong polarity of the 1π MO in CO. One can generalize here (also from the strong overlap in 3σ) to the conclusion that covalent bonding is only moderately reduced by a considerable amount of polarity in an MO. Even in the extreme case of LiF, the very appreciable overlap populations in the 3σ, 4σ, and 1π MOs (see Table 6) suggest that there is still a far from negligible amount of covalent bonding, although much of it may probably better be interpreted as mutual polarization of two ions. In any event, the bonding in LiF is *predominantly* ionic (see also the discussion of alkali halides in Section A, and of LiF in Section E). In such cases, also (but less so) in molecules such as BeO, the strong ionicity makes up, or more than makes up, in terms of bond dissociation energy, for the weakness of the covalent components of the bonding.

E. BONDING AND BINDING

Most of what was said on this subject in Section V.I.1 (q.v.) including Eqs. (1)–(4), is valid for heteropolar as well as homopolar diatomic molecules AB. $F_A = F_B$ ($=0$ at $R = \infty$ and at R_e) is still true, but now F_A [$= Z_B$, see Eq. (2) there] is no longer equal to f_B ($= Z_A$).

An extreme example of the heteropolar class is LiF, which Bader and Henneker [27] have discussed in detail in a paper entitled "The Ionic Bond." Using some good [28], but not the best, SCF data, they have computed a breakdown of the Hellmann–Feynman coefficients $f_{i\text{Li}}$ and $f_{i\text{F}}$ for the several MOs, and their totals f_{Li} and f_{F}. These are reproduced in Tables 15 and 16.

Tables 15 and 16 for the nearly ionic LiF may be compared and contrasted with Table 19 for the homopolar O_2 in Section V.I.1 (see also Table 20 there). As compared with two separate atoms Li and F, there is some polarization in 1σ and 2σ and especially 5σ (cf. Table 2 for LCSTF coefficients). However, the greatest effect is in 4σ, where $f_{i\text{Li}}^{\text{FF}}$ and $f_{i\text{F}}^{\text{FF}}$ values correspond to an almost complete transfer out of $2s_{\text{Li}}$ into $2p\sigma_{\text{F}}$. Nevertheless, $2p\sigma_{\text{F}}$ in 4σ is strongly polarized toward the Li (now nearly Li^+), as shown especially by the large negative $f_{i\text{F}}^{\text{FF}}$ polarization term and the large overlap

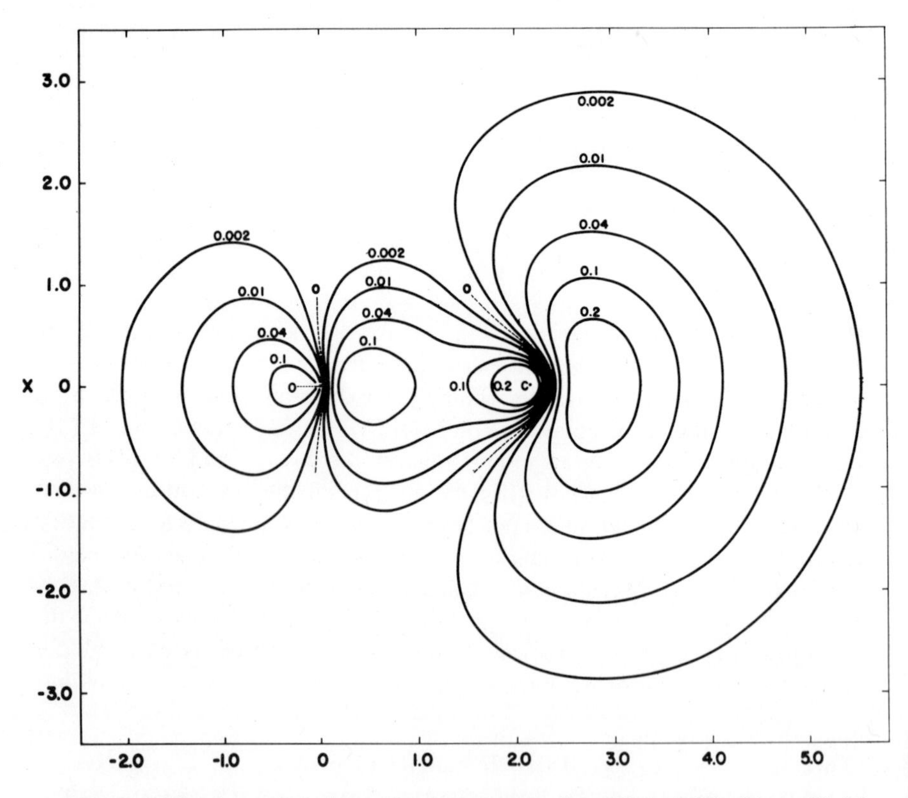

FIG. 5 Molecular orbital charge density contours; 5σ orbital of CO. [From W. M. Huo, *J. Chem. Phys.* **43**, 624 (1964).]

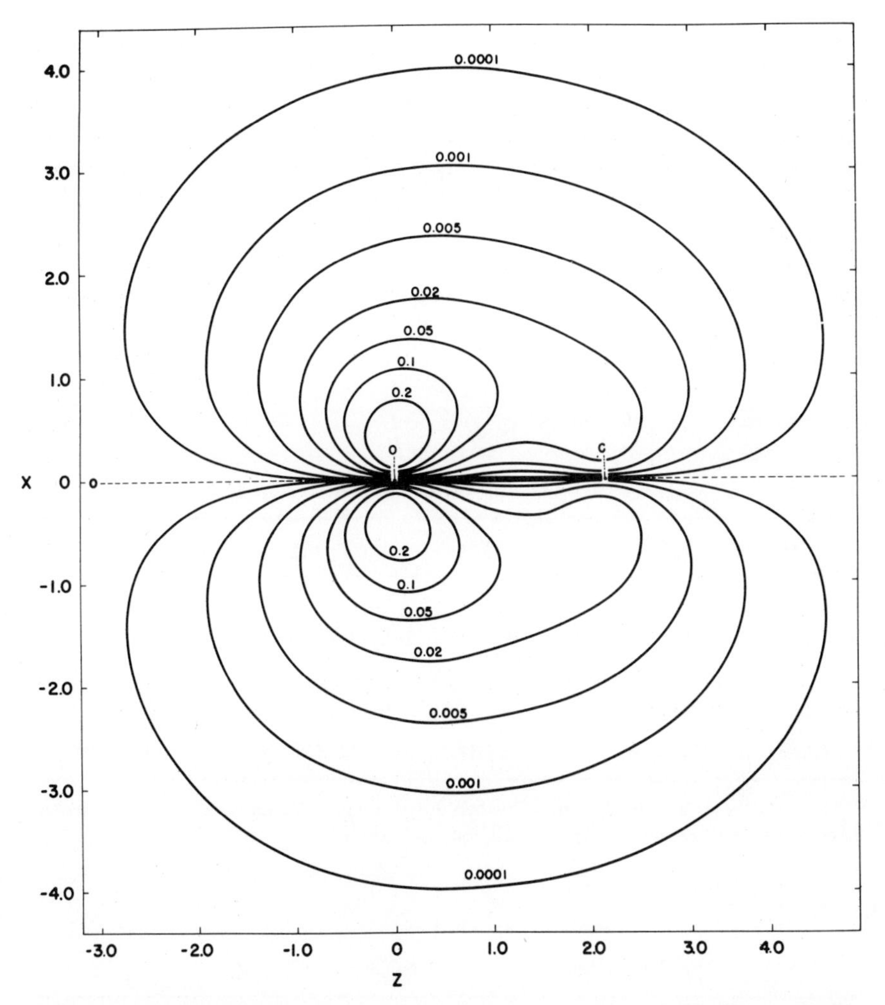

FIG. 6 Molecular orbital charge density contours; 1π orbital of CO. [From W. M. Huo, *J. Chem. Phys.* **43**, 624 (1964).]

term $f_{i\mathrm{F}}^{\mathrm{LiF}}$. The overlap population in 4σ is located much closer to the F than to the Li, as is shown by the much larger value of $f_{i\mathrm{F}}^{\mathrm{LiF}}$ than of $f_{i\mathrm{Li}}^{\mathrm{LiF}}$. On the other hand, the very appreciable overlap population in 1π is much more nearly equally shared by the two nuclei, as shown by the roughly equal values of $f_{i\mathrm{Li}}^{\mathrm{LiF}}$ and $f_{i\mathrm{F}}^{\mathrm{LiF}}$.

On the whole, the Hellmann–Feynman analysis as well as the charge distribution in Fig. 2 indicate that the LiF molecule is much closer to $\mathrm{Li^+F^-}$ than to $\mathrm{Li \cdot F}$ in structure, although with somewhat polarized ions. The increase in the screening coefficient $f_{i\mathrm{Li}}^{\mathrm{FF}}$ from 9 (for Li + F at large R) to 9.511 at R_e indicates a net transfer of charge from Li to F. The fact that the screening coefficient is less than 10 (as it would be for $Li^+ + F^-$ at large R) is due primarily to the ineffective screening of the π density and not to a large transfer of charge to the overlap region. The total overlap contribution $f_{\mathrm{Li}}^{\mathrm{LiF}}$ is in fact small. In covalent bonds, overlap contributions are large (see Chapter V).

Diatomic hydrides form a class of heteropolar molecules for which Bader *et al.* [26] have discussed bonding and binding in terms of a Hellmann–Feynman analysis. They classify LiH as ionic, but the other first-row hydrides as covalent, except for BeH which they consider intermediate.

TABLE 15

Force Coefficients for the Li Nucleus in LiF at R_e [a]

MO	$f_{i\mathrm{Li}}^{\mathrm{LiLi}}$	$f_{i\mathrm{Li}}^{\mathrm{LiF}}$	$f_{i\mathrm{Li}}^{\mathrm{FF}}$	$f_{i\mathrm{Li}}$
1σ	0.000 (0)	0.000 (0)	2.000 (2)	2.000 (2)
2σ	−0.362 (0)	0.070 (0)	0.003 (0)	−0.289 (0)
3σ	−0.064 (0)	0.016 (0)	2.004 (2)	1.956 (2)
4σ	−0.296 (0; 0)	0.080 (0)	2.200 (1; 2)	1.984 (1; 2)
1π	0.010 (0)	0.180 (0)	3.304 (4)	3.494 (4)
Totals:	−0.712 (0; 0)	0.346 (0)	9.511 (9; 10)	f_{Li} = 9.145 (9; 10)

[a] In parentheses are the f_{iA} and f_{iB} values for the separate atoms ($R \to \infty$); or (*second* entry for 4σ and Totals) for the separate ions Li^+ and F^-.

TABLE 16

Force Coefficients on the F Nucleus in LiF at R_e [a]

MO	$f_{i\mathrm{F}}^{\mathrm{FF}}$	$f_{i\mathrm{F}}^{\mathrm{LiF}}$	$f_{i\mathrm{F}}^{\mathrm{LiLi}}$	
1σ	0.099 (0)	0.001 (0)	0.000 (0)	0.100 (0)
2σ	0.007 (0)	0.028 (0)	1.972 (2)	2.007 (2)
3σ	0.636 (0)	0.081 (0)	0.014 (0)	0.731 (0)
4σ	−1.148 (0; 0)	0.739 (0)	0.043 (1; 0)	−0.366 (1; 0)
1π	0.474 (0)	0.136 (0)	0.012 (0)	0.622 (0)
Totals:	0.068 (0; 0)	0.985 (0)	2.041 (3; 2)	f_{F} = 3.094 (3; 2)

[a] In parentheses are the f_{iA} and f_{iB} values for the separate atoms ($R \to \infty$); or (*second* entry for 4σ and Totals) for the separate ions Li^+ and F^-.

F. CORRELATED WAVE FUNCTIONS; DIPOLE MOMENTS

The effects of electron correlation in heteropolar molecules have already been discussed somewhat extensively in Section IV.D and IV.E. (q.v.) for the case of diatomic hydrides. Table IV-17 shows that dipole moments computed with electron correlation are systematically smaller in magnitude and much closer to experiment than SCF-computed dipole moments. However, as was mentioned in Section A, the discrepancies are smaller for ionic molecules than for others. In Section IV.D it was pointed out that the improvements effected by CM calculations can be explained almost entirely by single MO substitutions on the SCF wave function.

In the case of CO, the SCF-computed dipole moment μ is nearly zero, which numerically is in as good agreement with the observed μ as in other cases. However, several writers have devoted much effort to explaining the small numerical disagreement, because they have been fascinated by the fact that the SCF-computed polarity is C^+O^-, whereas experiment makes it C^-O^+ ($\mu = -0.112$ D) [1b–d]. Some time ago, Mulliken [29] explained the smallness of the dipole moment of CO in terms of a competition between C^+O^- and C^-O^+ valence-bond structures whereby the moment has polarity C^-O^+ at moderately small and C^+O^- at moderately large R values (and becomes zero both as $R \to 0$ and as $R \to \infty$), and thus happens to pass through zero near R_e.

For the ground state of CO, Grimaldi *et al.* [30], using Huo's rather good basis set [1b] and adding CM up to 138 of the most important double substitutions plus *all* 62 single substitutions (see his Table III for a list of substitutions) obtained a μ of -0.08 D, of the correct sign and in close agreement with experiment. They noted that the correct result could not be obtained except by including the single substitutions, but only in combination with an ample selection of double substitutions. Further, they reported that the only important single substitutions are of the type $1\pi \to n\pi$ ($n =$ 2 to 8). Green [31] repeated the computations with slight variations and obtained similar results. With 201 double substitutions and no singles, Grimaldi *et al.* obtained 42% of the CE (correlation energy) but the wrong sign of μ; with 138 doubles and 62 singles, 39% of the CE but the correct sign of μ. Green with 117 doubles and 36 singles obtained 39% of the CE and $\mu = -0.12$ D. The need for single substitutions here should be compared with what was said in Sections IV.D and IV.E, with particular reference to OH.

In addition to his work on the ground state of CO, Green has made CM computations on the energy and dipole moment of the a $^3\Pi$ excited state, and of the A $^1\Pi$ state of CS, obtaining good agreements with experiment for the dipole moments [32].

Billingsley and Krauss have made an MCSCF study of the ground state of CO and its dipole moment function [33]. Their MCSCF function includes all CSFs necessary to give correct dissociation of the CO (to normal SCF atoms). They also give an extensive general discussion of the choice and classification of CSFs for use in MCSCF and CM; see Section IV.D for a review. In their MCSCF study, they use only double substitutions (relative to the single SCF dominant configuration), and point out that the re-optimization of the MOs which takes place during the MCSCF procedure has the same effect as the inclusion of single substitutions. (They use Huo's somewhat limited basis set [1b], as did Grimaldi *et al.* and Green.) A comparison of their MCSCF MOs (see their Table III) with those of Huo's SCF calculations, as reproduced here in our Table 1, shows considerable changes in the MCSCF versus the SCF MOs. In particular, the center of gravity of the 1π MO has been shifted somewhat from the O toward the C atom; there are smaller shifts in the σ MOs. (For a visualization of similar changes in an MCSCF treatment of OH, see Fig. IV.6.) These shifts together with some shifts in other correlation MOs explain why the computed dipole moment is changed from C^+O^- as in SCF to C^-O^+ in the MCSCF calculation. For increased accuracy a total of 11 CM terms is included in their MCSCF, yielding a computed μ of -0.167 D (see their Table I for a list of the CSFs which they include).

Billingsley and Krauss also compute and discuss the variation of μ with R near R_e. This variation is nearly linear in $R-R_e$ and is relevant to the transition moments of the infrared bands of CO [29]. Their computed variation agrees moderately well with what is obtained from the infrared experimental data.

The most accurate computation of the CE of the ground state of CO is by Siu and Davidson [34]. They note that the CE of CO is about 0.525 a.u. and that of its atoms about 0.409, so the CE contribution to the binding (the MECE) is about -0.12 a.u. Their best CM calculation yields about 70% of this MECE. With a basis set somewhat similar to Huo's but with additional high-l terms, and a total of 2484 CSFs, they obtain a computed μ of -0.33 D, and an energy $E=-113.1456$ a.u. Their Table V lists the 50 most important CSFs used. In addition to this main result, they discuss the summation of pair energies (single-shell pair substitutions only) and find a CE 93% of the experimental—too large because of the omission of partially counterbalancing split-shell and other substitutions. They also discuss natural geminals and their occupation numbers in detail (cf. Sections I.F and I.G).

Using a minimal basis set optimized for the two atoms, plus maximal CM, O'Neil and Schaefer [35] have made computations at no fewer than nine R values on all the 72 states of CO which can dissociate to a ^3P, ^1D,

^1S, and ^5S carbon atom plus a ^3P, ^1D, or ^1S oxygen atom. The atomic 1s orbitals were, however, kept doubly occupied. In this way, potential curves for 72 valence-shell states were obtained, including all those known experimentally; nine bound states not known experimentally are predicted. This work is similar to that reported on the valence-shell states of C_2, N_2, and O_2 (see Section V.F), and of NH [36].

In earlier approximate work, making use of the virtual orbitals from SCF calculations, Lefebvre-Brion, Moser, and Nesbet computed vertical excitation energies for several of the lower valence-excited and Rydberg states of CO [37]. The method was similar to that discussed at the beginning of Section V.A, except that an extended instead of a minimal basis set was used. Thereby a considerable number of virtual MOs is obtained, the lowest of which correspond to excited MOs. For the Rydberg MOs, the extended basis set included STFs of relatively low ζ, which are needed if one wishes to approximate Rydberg MOs. For the remaining electrons, essentially a frozen core was used. After a calibration in terms of the empirical ionization potential of CO, good agreements with experimental excitation energies were obtained. However, a straightforward SCF treatment of much of the excited states, now feasible, would be more accurate. A paper by Rose and McKoy on some Σ states of CO, N_2, and O_2 should be mentioned here [38].

Heil and Schaefer [39] have made minimal basis plus maximal CM calculations on the 72 states of SiO derivable from silicon ^3P, ^1D, ^1S, and ^5S atoms and oxygen ^3P, ^1D, and ^1S atoms. The resulting potential curves strongly resemble those of CO.

Schaefer and Heil have made similar calculations on the lower excited states of CN, except

(a) they considered only those 59 states which dissociate to ^3P, ^1D, or ^1S carbon atoms and ^4S or ^2D nitrogen atoms;

(b) the 2s and 2p STFs used were optimized for the molecule [40].

As before, the K shell MOs were kept doubly occupied. Nine bound states not experimentally known are predicted, including some low-energy quartet states. Most of the states lie below the energies where Rydberg states are expected, so relatively few cases of interaction with the latter are expected. NOs and their occupation numbers were computed for the lowest bound state of each symmetry.

For the X $^2\Sigma^+$ and B $^2\Sigma^+$ states of CN, using an extended basis set with moderate-sized CM, Green has computed the dipole moments with good agreement with experiment [41a]. Das *et al.* have made a careful calculation on the X $^2\Sigma^+$ and five valence-excited states of CN [41b]. They have computed spectroscopic constants and potential curves for all the states and dipole moments at R_e for the X $^2\Sigma^+$ and A $^2\Pi$ states (1.48 D, C^+N^-

for the former, in agreement with the experimental 1.45 D, and 0.314 D, C^-N^+ for the latter).

Kouba and Öhrn have made a calculation rather similar to that of Schaefer and Heil on CN for 54 low-lying valence-shell states of BC [42]. They show that the ground state must be $^4\Sigma^-$.

Bagus and Preston [43] made SCF and limited CM calculations on the $^5\Sigma^+$ and related states of FeO, and very extended CM calculations on the lowest $^3\Sigma^+$ state, which, however, they conclude is not the ground state.

For the $1\sigma^2 2\sigma^2 3\sigma^2 4\sigma^2 1\pi^3$, $^2\Pi$ ground state and the $1\sigma^2 2\sigma^2 3\sigma^2 4\sigma 1\pi^4$, $^2\Sigma^+$ state of LiO, Yoshimine, using an extensive basis set, has obtained potential curves by CM calculations [44]. The K shells are not correlated. Yoshimine, McLean, and Liu have used these and similar calculations on AlO to compute rather accurate band strengths for electric dipole transitions involving the low-energy states of these molecules [45]. They give a useful review of transition moment theory.

In the $\ldots 5\sigma^2 6\sigma^2 7\sigma 2\pi^4$, $^2\Sigma^+$ ground state of AlO, SCF calculations give *two* different solutions, a situation similar to that for BeH as reported in Section IV.A. On introducing CM and making use of NOs, a single potential curve is obtained. The CM function finally presented contains 1756 CSFs [45]. Das *et al.* have made a careful MCSCF calculation on the X $^2\Sigma^+$ and A $^2\Pi$ states, and on the oscillator strength, as a function of R, for the transition between them [41b]. They compare their results with those of Yoshimine *et al.*

Bertoncini *et al.* [46] have made OVC calculations at many R values and obtained potential curves for the ground states $\ldots 5\sigma^2$, $^1\Sigma^+$ of NaLi, using CM with $6\sigma^2$, $\ldots 2\pi^2$, and $\ldots 7\sigma^2$ to give left–right, angular, and in–out correlation, and correct dissociation. Complete details are given on the forms of all including the core MOs, and contour diagrams are shown for the 5σ, 6σ, 2π, and 7σ MOs, also total densities and difference densities relative to the separate atoms. SCF calculations and corresponding potential curves are also presented for the $\ldots 5\sigma 2\pi$, $^3\Pi$ and $^1\Pi$ and the (repulsive) $\ldots 5\sigma 6\sigma$, $^3\Sigma^+$ states and for the $\ldots 5\sigma$, $^2\Sigma^+$ ground state of $NaLi^+$. As in the case of other alkali metal molecules (e.g., Li_2^+ versus Li_2), the dissociation energy is greater for the positive ion than for the neutral molecule. In the case of $\ldots 5\sigma 6\sigma$, $^3\Sigma^+$ the SCF calculations without CM give a potential curve with correct dissociation (cf. $1\sigma_g 1\sigma_u$, $^3\Sigma^+$ of H_2—see Section III.B). Curves of potential energy (V) and kinetic energy (T) as a function of R are given for the $^1\Sigma^+$, $^3\Pi$, and $^2\Sigma^+$ states. As is typical for stable states (cf. Section II.F) V at first drops and T rises at large R during molecular formation, while at smaller R these relations are reversed.

Rosmus and Meyer [47] have made new calculations by the CEPA (see Section I.G) method on the spectroscopic constants and dipole moment functions for the ground state of NaLi. Previous calculations [46] give surprisingly poor agreements with experiment for these, especially for the dipole moment (1.24 D and 0.99 D, experiment 0.46 ± 0.01 D). Previous investigations included only valence-shell correlation energy. In work on LiH and NaH, Rosmus and Meyer found that for these one-valence-electron hydrides, core-valence-shell contributions to the correlation energy are usually important for the above-mentioned properties. They have now found that the same is true for NaLi; on including the core-valence contributions, their computed dipole moment is 0.485 D. The agreements for R_e, ω_e, and $x_e\omega_e$ are also very considerably improved.

Liu and Schaefer using an extensive basis set including d and f functions have made accurate nonrelativistic SCF calculations on the ground state of KrF (repulsive) and KrF^+ (stable) [48]. CM calculations using 158 configurations for KrF and 210 for KrF^+ have also been made. Potential curves and computed properties are discussed. Dunning and Hay [49a] using the POL–CI method with very extensive CM have obtained potential curves also for the bound excited states of KrF which arise from Kr^+ plus F^- and from Kr^* (^3P) plus F. One of the resulting states is responsible for the observed laser transition. They have also investigated the rare gas oxides [49b].

In order to resolve the question as to whether the ground state of BeO is $\ldots 4\sigma^2 1\pi^4$, $^1\Sigma^+$ or $4\sigma 1\pi^3 5\sigma$, $^3\Pi$, Schaefer *et al.* have made rather extensive CM calculations on the two states [50]. Although in SCF calculations the $^3\Pi$ is lower, with CM the $^3\Pi$ is found to be 0.73 eV above the $^1\Sigma^+$ state. Still higher are found $\ldots 4\sigma^2 1\pi^3 2\pi$, $^3\Sigma^+$ and (repulsive) $4\sigma^2 1\pi^2 5\sigma^2$, $^3\Sigma^-$. NO occupation numbers and coefficients are given. Analogies to the isoelectronic molecule C_2, where the ground state is $^1\Sigma^+$ only by a very small margin, are incomplete. Earlier, Verhaegen *et al.* [51] made SCF calculations on the likewise isoelectronic molecule BN and after estimating relative correlation energies of various states, concluded that the ground state is $^3\Pi$. This conclusion is supported by the observation for BN of triplet spectra, but perhaps should not be considered final.

For the NO molecule, a great number of Rydberg states are known, many of them perturbed by valence-shell states [52]. Lefebvre-Brion and Moser have made SCF computations on many of these Rydberg states, also on some states of BF [53].

A limited CM calculation with a somewhat limited SCF basis set has been made for the $\ldots 5\sigma^2 1\pi^4 2\pi$, $^2\Pi$ ground state of NO at its R_e by Kouba and Öhrn [54]. A natural orbital iteration procedure was used (see Section I.F). After making the best SCF calculation to date for $^2\Pi$ at R_e, Green made a large CM calculation using 20σ and 12π STFs in the basis set; he has also calculated the $\ldots 5\sigma^2 1\pi^4 6\sigma$, A $^2\Sigma^+$ Rydberg state in a similar way [55]. Green finds good agreement with experiment for the dipole moment of the $^2\Pi$, which is shown to be of polarity N^-O^+, but poorer agreement for the $^2\Sigma^+$. Other computed properties are also presented.

Walsh and Goddard recomputed the dipole moment of the A $^2\Sigma^+$ state and obtained good agreement with experiment [56]. They also computed the dipole moments and some other properties of the ground state and the D $^2\Sigma^+$ Rydberg state.

Using Green's basis set, Billingsley [57] has made 20-term MCSCF (OVC) plus CM calculations on NO $^2\Pi$ over a range of R values from 1.6 to 3.4 a.u. (R_e is at 2.175 a.u.), and has computed a potential curve and a dipole moment curve over this range. The potential curve agrees closely with experiment, and the computed dipole moment of -0.138 D (N^-O^+) for the $v = 0$ vibration level is in close agreement with the experimental value of ± 0.158 D. The computed dipole moment function passes through zero at about 2.3 a.u., and will be used to compute infrared transition probabilities. The behavior of the dipole moment as a function of R is similar to that of CO (see earlier discussion in this section).

An interesting feature of the calculations is that at about 3.1 a.u., not far beyond R_e, the configuration $\ldots 5\sigma^2 1\pi^3 2\pi^2$ becomes equally as important as $\ldots 5\sigma^2 1\pi^4 2\pi$, which is the main configuration of the B $^2\Pi$ state at its R_e. Values of a number of one-electron properties as a function as R are computed and tabulated.

Thulstrup *et al.* have computed potential curves using a minimal basis set plus CM for a number of the low-energy states of NO ($^2\Pi$ and $^4\Pi$ states), NO^+ ($^1\Sigma^+$ and triplet states), NO^- ($^3\Sigma$ states) and NO^{2+} ($^2\Sigma$, $^2\Pi$, and $^2\Delta$ states) [58]. Experimentally known states, including NO^+ states from photoelectron spectroscopy, are in harmony with the results of the calculations.

Bagus and Schaefer [59a][5] have reported SCF calculations on the $^3\Pi$ and $^1\Pi$ 1s hole states of NO^+ (configurations $1s \ldots 1\pi^4 2\pi$, $^{1,3}\Pi$). Agreement with experiment is good, for both the N atom and the O atom 1s holes, when the NO and NO^+ states are separately computed and the differences compared with the observed ionization potentials. Evidently the corrections needed because of the differences in CM between NO and NO^+ are very small here. Further, the computed singlet–triplet intervals agree with experiment.

Schaefer *et al.* [60] have made SCF plus CM calculations showing that the ground state of LiN is $^3\Sigma^-$, with $^3\Pi$ a little higher. They have calculated potential curves for these states, and the spectroscopic transition moment between them.

Many molecules on electron impact show "resonances" that correspond

[5] Compare also Bagus *et al.* [59b] on some intensity questions.

to states of their negative ions. For NO there are several sharp resonances which Lefebvre-Brion [61] in a CM calculation has shown to be attributable to formation of NO^- states with ...$(Rs\sigma)^2$, $^1\Sigma^+$, $(Rs\sigma)(Rp\pi)$, $^3\Pi$, $(Rs\sigma)(Rp\sigma)$, $^3\Sigma^+$, and $(Rp\pi)^2$, $^3\Sigma^-$; where $Rs\sigma$ and $Rp\pi$ are Rydberg MOs. She used a double-ζ basis set (21σ and 9π STFs) for the K and L shells, plus two δ AOs for the CM; CM was used only for the Rydberg MOs.

Although not strictly related to this Section, the use of the equations-of-motion method to compute excitation energies, potential curves, and oscillator strengths of spectroscopic transitions deserves mention [62]. Computations have been made on CO and N_2. Excitation energies are computed to within 10% of experiment and mostly better, and oscillator strengths agree well with experiment. The method is economical in computation time.

REFERENCES

1. (a) P. E. Cade, W. M. Huo, and J. B. Greenshields, *Atomic Data* **15**, 1 (1975).
   (b) W. M. Huo, *J. Chem. Phys.* **43**, 624 (1965).
   (c) A. D. McLean and M. Yoshimine, *Int. J. Quantum Chem.* **15**, 313 (1967); and in detail, *IBM J. Res. Develop. Suppl.* (1967).
   (d) S. Green, *J. Chem. Phys.* **52**, 3100 (1970).
2. G. M. Schwenzer, D. H. Liskow, H. F. Schaefer, III, P. S. Bagus, B. Liu, A. D. McLean, and M. Yoshimine, *J. Chem. Phys.* **58**, 3181 (1973).
3. (a) R. L. Matcha, *J. Chem. Phys.* **47**, 4595, 5295 (1967); **48**, 335 (1967); **49**, 1264 (1968); **53**, 485 (1970).
   (b) R. L. Matcha, *ibid.* **65**, 1962 (1976); *J. Am. Chem. Soc.* **95**, 7505 (1973).
4. G. Doggett and A. M. Kendrick, *J. Chem. Soc.* **A6**, 825 (1970).
5. L. A. Curtiss, C. W. Kern, and R. L. Matcha, *J. Chem. Phys.* **63**, 1621 (1975).)
6. K. D. Carlson and K. R. Nesbet, *J. Chem. Phys.* **41**, 1051 (1964); on TiO; K. D. Carlson, E. Ludena, and C. Moser, *ibid.* **43**, 2408 (1965); K. D. Carlson and C. Moser, *J. Phys. Chem.* **67**, 2644 (1963); on TiO; *J. Chem. Phys.* **44**, 3259 (1966); on VO; *ibid.* **46**, 35 (1967), on ScF; K. D. Carlson, C. R. Clayton, and C. Moser, *ibid.* **46**, 4963 (1967), on TiN; I. Cohen and K. D. Carlson, *J. Phys. Chem.* **73**, 1356 (1969) on density distributions and bonding in TiO and ScF; for some further SCF calculations on ScF, see P. R. Scott and W. G. Richards, *Chem. Phys. Lett.* **28**, 101 (1974).
7. P. A. G. O'Hare and A. C. Wahl, *J. Chem. Phys.* **53**, 2469 (1970); OF and its ions; *ibid.* **53**, 2834 (1970), SF and SeF and their ions, *ibid.* **54**, 3770 (1971), ClO and its ions; *ibid.* **54**, 4563 (1971), NF, PF, and their ions; *ibid.* 55, 666 (1971), CF and SiF and their ions, P. A. G. O'Hare; *ibid.* **59**, 3842 (1973), NF, PF, SiF, SF; *ibid.* **60**, 4084 (1974), SeF; P. A. G. O'Hare, A. Batana, and A. C. Wahl, *ibid.* **59**, 6495 (1973), AsF.
8. J. J. Kaufman and L. M. Sachs, *J. Chem. Phys.* **52**, 645, 3534 (1970); J. J. Kaufman, *ibid.* **58**, 1680, 440 (1973).
9. P. A. G. O'Hare and A. C. Wahl, *J. Chem. Phys.* **56**, 4516 (1972); NaO, NaO^+, NaO^-; S. P. So and W. G. Richards, *Chem. Phys. Lett.* **32**, 227 (1975); NaO, KO, and RbO.
10. G. Verhaegen, W. G. Richards, and C. M. Moser, *J. Chem. Phys.* **46**, 160 (1967); see also M. P. Melrose and D. Russell, *ibid.* **55**, 470 (1971); **57**, 2586 (1972).
11. W. M. Huo, K. F. Freed, and W. Klemperer, *J. Chem. Phys.* **46**, 3556 (1967).
12. J. Schamps and H. Lefebvre-Brion, *J. Chem. Phys.* **56**, 573 (1971).

13. D. Carlson, K. Kaiser, C. Moser, and A. C. Wahl, *J. Chem. Phys.* **52**, 4678 (1970).
14. M. Yoshimine, *J. Chem. Phys.* **40**, 2970 (1974), on BeO; *J. Phys. Soc. Japan* **25**, 1100 (1968).
15. J. Cambray, J. Gasteiger, A. Streitwieser, Jr., and P. S. Bagus, *J. Amer. Chem. Soc.* **96**, 5978 (1974).
16. H. Lefebvre-Brion and C. M. Moser, *Phys. Rev.* **118**, 675 (1960); C. C. Lin, K. Hijikata, and M. Sakamoto, *J. Chem. Phys.* **33**, 878 (1960).
17. C. Jungen and H. Lefebvre-Brion, *J. Mol. Spectros.* **33**, 520 (1970).
18. D. H. Boyd, *J. Chem. Phys.* **52**, 4846 (1970).
19. (a) R. S. Mulliken, *Rev. Mod. Phys.* **4**, 1 (1932), Section C6.
    (b) R. S. Mulliken, *Phys. Rev.* **38**, 836 (1931).
20. T. E. H. Wlaker and W. G. Richards, *Proc. Phys. Soc. London* **92**, 285–290 (1967).
21. T. E. H. Walker and W. G. Richards, *J. Phys. B.* **3**, 271 (1970).
22. T. E. H. Walker and W. G. Richards, *J. Phys. B.* **1**, 1061 (1968).
23. (a) T. E. H. Walker and W. G. Richards, *Symp. Faraday. Soc* **2**, 64 (1968); *Phys. Rev.* **177**, 100 (1969); *J. Chem. Phys.* **52**, 1311 (1970).
    (b) R. K. Hinkley, J. A. Hall, T. E. H. Walker and W. G. Richards, *J. Phys. B.* **5**, 204 (1972).
24. R. F. W. Bader and A. D. Bandrauk, *J. Chem. Phys.* **49**, 1653 (1968).
25. R. F. W. Bader, W. H. Henneker, and P. E. Cade, *J. Chem. Phys.* **46**, 3341 (1967).
26. R. F. W. Bader, I. Keaveny, and P. E. Cade, *J. Chem. Phys. Phys.* **47**, 3381 (1967).
27. R. F. W. Bader and W. H. Henneker, *J. Am. Chem. Soc.* **87**, 3063 (1965).
28. A. D. McLean, *J. Chem. Phys.* **39**, 2653 (1963).
29. R. S. Mulliken, *J. Chem. Phys.* **2**, 400, 712 (1934).
30. F. Grimaldi, A. Lecourt, and C. Moser, *Int. J. Quantum Chem.* **1S**, 153 (1967).
31. S. Green, *J. Chem. Phys.* **54**, 827 (1971).
32. S. Green, *J. Chem. Phys.* **56**, 739 (1972); **57**, 2830 (1972).
33. F. P. Billingsley, II, and M. Krauss, *J. Chem. Phys.* **60**, 4130 (1974).
34. A. K. Q. Siu and E. R. Davidson, *Int. J. Quantum Chem.* **4**, 223 (1970).
35. S. V. O'Neil and H. F. Schaefer, III, *J. Chem. Phys.* **53**, 3994 (1970).
36. J. Kouba and Y. Öhrn, *J. Chem. Phys.* **52**, 5387 (1970).
37. H. Lefebvre-Brion, C. M. Moser, and R. Nesbet, *J. Chem. Phys.* **35**, 1702 (1961); *J. Mol. Spectros.* **13**, 418 (1964).
38. J. B. Rose and V. McKoy, *J. Chem. Phys.* **55**, 5435 (1971).
39. T. G. Heil and H. F. Schaefer, III, *J. Chem. Phys.* **56**, 958 (1972).
40. H. F. Schaefer, III, and T. G. Heil, *J. Chem. Phys.* **54**, 2573 (1971).
41. (a) S. Green, *J. Chem. Phys.* **57**, 4694 (1972).
    (b) G. Das, T. Janis, and A. C. Wahl, *ibid.* **61**, 1274 (1974).
42. J. E. Kouba and Y. Öhrn, *J. Chem. Phys.* **53**, 3923 (1970).
43. P. S. Bagus and H. J. T. Preston, *J. Chem. Phys.* **59**, 2986 (1973).
44. M. Yoshimine, *J. Chem. Phys.* **57**, 1108 (1972).
45. M. Yoshimine, A. D. McLean, and B. Liu, *J. Chem. Phys.* **58**, 4412 (1973).
46. (a) P. J. Bertoncini, G. Das, and A. C. Wahl, *J. Chem. Phys.* **52**, 5112 (1972).
    (b) S. Green, *ibid.* **54**, 827 (1971).
47. P. Rosmus and W. Meyer, *J. Chem. Phys.* **65**, 492 (1976).
48. B. Liu and H. F. Schaefer, III, *J. Chem. Phys.* **55**, 2369 (1971).
49. (a) T. H. Dunning, Jr., and P. J. Hay, *J. Chem. Phys.* **66**, 1306 (1977).
    (b) T. H. Dunning, Jr., and P. J. Hay, *J. Chem. Phys.* **66**, 3767 (1977).
50. H. F. Schaefer, III, *J. Chem. Phys.* **55**, 176 (1971); S. V. O'Neil, P. K. Pearson, and H. F. Schaefer, III, *Chem. Phys. Lett.* **10**, 404 (1971); *J. Chem. Phys.* **56**, 3938 (1972).

51. G. Verhaegen, W. G. Richards, and C. M. Moser, *J. Chem. Phys.* **46**, 160 (1967).
52. Cf. K. Dressler and E. Miescher, *Astrophys. J.* **141**, 1266 (1965), and references therein.
53. H. Lefebvre-Brion and C. Moser, *J. Mol. Spectros.* **15**, 211 (1965); *J. Chem. Phys.* **44**, 2951 (1966).
54. J. E. Kouba and Y. Öhrn, *Int. J. Quantum Chem.* **5**, 539 (1971).
55. S. Green, *Chem. Phys. Lett.* **13**, 552 (1972); **23**, 115 (1973).
56. S. P. Walsh and W. A. Goddard, III, *Chem. Phys. Lett.* **33**, 18 (1975).
57. F. P. Billingsley, II, *J. Chem. Phys.* **62**, 864 (1975).
58. E. W. Thulstrup and Y. Öhrn, *J. Chem. Phys.* **57**, 3716 (1972); P. W. Thulstrup, E. W. Thulstrup, A. Andersen, and Y. Öhrn, *ibid.* **60**, 3975 (1974).
59. (a) P. S. Bagus and H. F. Schaefer, III, *J. Chem. Phys.* **55**, 1474 (1971).)
    (b) P. S. Bagus, M. Schrenk, D. W. Davis, and D. A. Shirley, *Phys. Rev.* **9A**, 1090 (1974).
60. C. E. Dykstra, P. K. Pearson, and H. F. Schaefer, III, *J. Am. Chem. Soc.* **97**, 2321 (1975).
61. H. Lefebvre-Brion, *Chem. Phys. Lett.* **19**, 456 (1973).
62. T. Shibaya and V. McKoy, *Phys. Rev.* **A2**, 2208 (1970); J. Rose, T. Shibaya, and V. McKoy *J. Chem. Phys.* **58**, 74 (1973); W. Coughran, J. Rose, T. Shibaya, and V. McKoy, *ibid.* **58**, 2699 (1973).

INDEX

I

K

L

A
B 7
C 8
D 9
E 0
F 1
G 2
H 3
I 4
J 5